The World's Worst Warships

ANTONY PRESTON

Conway Maritime Press

© Antony Preston 2002.

All rights reserved. No part of this book may be reproduced or transmitted in any form without prior written permission from the publisher.

This edition published in Great Britain in 2002 by
Conway Maritime Press,
64, Brewery Road,
London, N7 9NT.
www.conwaymaritime.com

A member of Chrysalis Books plc

9 8 7 6 5 4 3 2 1

A CIP catalogue record for this book is available from the British Library.

ISBN 0 85177 754 6

Set in 9/12 Bembo.

All illustrations courtesy of Chrysalis Images.

Printed and bound in Spain.

DEDICATION

TO MY GRANDSON JONATHAN

Contents

Introduction	9
Civil War monitors	16
Turret Ship HMS *Captain*	21
Vitse-Admiral Popov and *Novgorod* Coast Defence Ships	26
Armoured rams HMS *Polyphemus* and USS *Katahdin*	30
Armoured cruiser *Rurik*	35
Dynamite cruiser USS *Vesuvius*	40
Powerful class protected cruisers	44
Borodino class battleships	50
Destroyer HMS *Swift*	56
Viribus Unitis class dreadnoughts	61
Normandie class dreadnoughts	67
AA class fleet submarines	73
'Flush-decker' destroyers	78
K class submarines	84
HM Ships *Courageous, Glorious, Furious,* light battlecruisers	90
Fast battleships HMS *Hood*	96
Omaha class scout cruisers	102
HMSwS *Gotland* hybrid cruiser	108
Duquesne class heavy cruisers	112
Deutschland class 'pocket battleships'	117
Condottieri class light cruisers	123
IJNS *Ryujo* aircraft carrier	130
Mogami class cruisers	136
Yamato class super battleships	142
Bismarck class battleships	148
Implacable class fleet aircraft carriers	153
Hydrogen-Peroxide Submarines	159
Alpha class nuclear attack submarines	165
Type 21 anti-submarine frigates	171
La Combattante-type fast attack craft	177
Glossary	184
Bibliography	185
Index	187

ACKNOWLEDGEMENTS

I wish to thank all those friends and fellow seekers after wisdom and truth who either contributed memories, constructive criticism or have written elsewhere on related aspects of design and performance. This 'band of brothers' includes the late David Lyon and the late George Osbon (former colleagues at the National Maritime Museum), Bob Todd (currently working at the Museum's Brass Foundry out-station), Dr Norman Friedman, Stuart Slade, Bradley Perrett, John Roberts (a long-standing friend and collaborator), the late David (J D) Brown (former head of the Royal Naval Historical Branch), David K Brown RCNC, the late Ian Hogg (who educated an entire generation on the science of artillery and guns), the late John Campbell (for his wide-ranging and deep knowledge of the Edwardian Era), Andrew Gordon (whose account of the Battle of Jutland has set a new benchmark), and all those who have contributed in one way or another, but cannot list in full. My thanks also go to my long-suffering publishers, for their patience and forbearance during what turned out to be a much harder slog than I had anticipated.

Antony Preston
London May 2002

INTRODUCTION

Although the subject of warship-design might be dismissed as a pastime for professionals and an even smaller nerdish audience, this reflects nothing more than ignorance among many 'mainstream' naval historians. Yet ever since navies became formal national organisations the quality of ships has been a major factor in both their successes and their failures. To those of us who study navies, one of the fascinating aspects is the complexity of the subject, which encompasses engineering, economics, sociology and even psychology.

Landsmen are mostly unaware of the incontrovertible fact that a ship is the largest mobile structure on the planet. Warships take the story even further, because they combine the complex engineering inherent in shipbuilding with the unavoidable fact that they are intended to fight other ships and to take hits from their opponents. Because naval architects are often unwilling to discuss their profession (and naval designers are usually required to be very discreet) their key role in the achievement of designing and building a ship is not common knowledge. The result is a serious lack of understanding on the part of most of the public and the media.

An example of this in recent years was the aftermath of the Falklands conflict in 1982. Many commentators declared that the big warship (by which they meant the destroyer and the frigate) was no longer worth building. In the United Kingdom a lobby appeared from nowhere, deriding Royal Navy designs and promoting a 'novel' hullform, on the grounds that Navy designers were wedded to a 'long, thin' hull as some sort of obsolete tradition. The ensuing rumpus was seized on by the British media, which portrayed the argument as some sort of David and Goliath contest. The allegedly hidebound Royal Corps of Naval Constructors (RCNC) was using its bureaucratic edge to muzzle a devoted band of heroes who would transform the Royal Navy if given money to build a prototype. But any attempt to inject a note of realism into the argument was always dismissed by those who knew better, claiming that the critics were 'in the pocket' of the Ministry of Defence. It was a brilliant example of targeted lobbying, creating a climate in which totally unqualified laymen attacked anyone not in favour of the 'breakthrough'. Ultimately it failed because it became clear to the Ministry of Defence and the Naval Staff that the design could never meet naval requirements, but as the Duke of Wellington said of Waterloo, 'it was a damned near-run thing'.

It was all too reminiscent of the tragic tale of HMS *Captain*, a subject dealt with later in the book, but there are many examples of the public's total inability to compare like with like. A famous ship of her day was the Chilean protected cruiser *Esmeralda*, which was in theory capable of delivering more firepower than a contemporary battleship! The *Esmeralda* was a very fine cruiser, but the statistics quoted were misleading. In this case it was very enthusiastic marketing by the British builder, which had nothing to gain by contradicting the publicity. Much closer to home was the mismatch between claims made in the West during the Cold War and the facts which emerged after the collapse of the Soviet Union. This time it was not the fault of the Soviets, who divulged virtually nothing about their ships, but misunderstandings among NATO naval officers and the naval intelligence community. A 2000-ton US Navy frigate would be compared unfavourably with a 9000-ton Soviet cruiser, for example, which should not have come as a surprise to professional seamen. Throughout the Cold War the US Navy was attacked for its alleged inability to build ships capable of matching their Soviet counterparts. As the distinguished technical historian Norman Friedman has pointed out, naval intelligence is all too frequently run by officers with no background in ship-design. The result was that all Soviet ships were credited with very high speeds and perfectly functioning weapon systems. When more knowledgeable commentators questioned the assumption that the Soviets had perfected technology not known to the West, they were told that the technical superiority stemmed from the communist system. I made myself unpopular at a seminar by suggesting that the implication of that explanation was that the West should adopt communism to allow fair competition. But any serious debate was made impossible by the knee-jerk response from the intelligence community, 'We have evidence but we can't say what it is.'

There are six basic factors which influence warship-design:

Cost
Perceived threats
Industrial capacity
Design competence
The operating environment
Incorrect post-battle analysis

Major warships are expensive capital investments, so cost will always be crucial. Many critics of Western navies during the Cold War assumed that the rising cost of warships was the result of 'gold plating'. The corollary was the assumption that the Soviet Navy had no financial constraints. In fact both assumptions were wrong. The rising level of danger from anti-ship missiles, new torpedo technology and other threats created a requirement for ships capable of surviving in a hostile environment. The Soviets did not publish defence estimates, nor did they indulge in public debate, but the expanding navy created by the drive of Admiral Sergei Gorshkov did consume vast resources. Somewhere within the Soviet military-industrial complex someone had to make difficult choices about priorities. Some external discipline must be applied, for the simple reason that any group of designers would wish to spend unlimited amounts on their pet projects. The customer, in this case the parent navy, also creates a problem: a demand for superiority to any equivalent, even those belonging to allies. 'Keeping up with the Jones's' is not confined to the consumer society; 'the Best is often the enemy of the Good'.

Closely linked to the problems of costs is the perception of the threat. This is as old as the hills, and in Nelson's time many naval officers were convinced that French men o'war were better than anything built for their own navy. It is still widely believed that the only positive influence was the superiority of French ships taken as prizes and taken into British service. Yet, objective analysis by modern historians has revealed that the Navy Board, which had responsibility for materiel, found that French ships could not stow sufficient stores for long periods at sea, the standard of construction was poor, and worst of all, they could not carry the weight of armament demanded by the Royal Navy. The poor standard of construction also meant that ex-French prizes needed much more money spent on repairs and maintenance. And if this is dismissed as a 'not invented here' attitude, look no further than French opinions of British warships; the standard complaint was that Royal Navy warships were, type for type, more heavily armed than their own.

It is true that a few ships such as the two-decker *Canopus,* captured at the Battle of the Nile, impressed everybody who knew her; and she was the forerunner of a series of large 2nd Rates. The fledgling United States' Navy (USN) built a few very big heavily-armed frigates, which performed very well in the War of 1812 against the small 28-gun frigates of the enemy, but as soon as the Royal Navy could spare more capable frigates from the European theatre, disasters such as the capture of the USS *Chesapeake* by HMS *Shannon* began to happen. By the end of that unnecessary war even such heroes as Stephen Decatur were forced to take refuge in coastal inlets to avoid capture by the blockaders.

There is another factor which is rarely touched on. After a single-ship action it is only human nature for a captain to magnify his achievement if he has defeated a more powerful opponent. Conversely, if defeated he has an excellent alibi—he was overwhelmed by a stronger opponent. There was another factor in the 18th century; a captain's prize, if taken into the Royal Navy yielded a very handsome financial bonus from the Admiralty. Is it surprising that ex-French prizes were praised to the skies? What admirers of France's imaginary lead over the Royal Navy are unaware of the Navy Board assessments of those prizes when docked for repairs and changes to be made. They contain endless complaints about shoddy construction and the expense of bringing the ship to the required Royal Navy standard. The French naval administration deserves high praise however, for making ship-design a science rather than merely repeating what had worked before.

This caveat about praising the opposition is valid for the 20th century as well, and could surface again in this century. The British beatification of the *Bismarck* (discussed in detail on pages 151-2) is a good example of the genre; and the absurd worship of Soviet designs merely proves that the vice was alive and well in the 1980s.

Which brings us conveniently to the malign role played by incorrect intelligence. Later in the book there are prize specimens such as the Russian cruiser *Rurik*, whose supposed superiority pushed the Royal Navy to build at great expense two enormous cruisers in reply. In 1917 intelligence reports of German 'super destroyers' led the Royal Navy to order a new powerful design as a counter. Exceptions prove the rule, and those V and W classes turned out to be the dreadnoughts of the destroyer scene, whose features were copied around the world and made a big contribution in the Second World War. Navies do best when they stick to building warships tailored to their own requirements; cloning an opponent's ships is never worthwhile, as a confrontation between the two rival designs can almost be guaranteed never to take place. As for stealing the blueprints of an opponent's ship, this a dead end because no group of engineers or designers would dream of doing a straight 'Chinese copy'. The normal comment on seeing a rival design seems to be, 'What a strange way to do things.'

It must never be forgotten that in peacetime military intelligence agencies are prone to lose objectivity. They are unwilling to share information in case someone else will spoil their game, and individuals often try to become experts on a specific threat to their enhance career prospects. Even unconsciously, they over-emphasise the threat to get extra funding for their navy's latest system, or they accept a 'mirror-image' of the enemy, trying to fit what they see into a framework based on their own side's tactics. There are few known examples of lying, but plenty of evidence of crediting the enemy with a non-existent margin of superiority. The corollary is to blame 'our' designers for not achieving what 'theirs' can. A minute dating from the 1930s is not untypical. The British Director of Naval Intelligence (DNI) attacked the Director of Naval Construction (DNC) for not matching the Japanese Furutaka class cruisers in speed, armament and compactness. To which the DNC retorted that the figures were either untrue (the correct explanation) or the Japanese had built the ships out of cardboard!

Technical and industrial capability have their own low-profile influence on the process. The reason why the Royal Navy ruled the roost for so long was the dominance of British industry. To the layman one 100-gun ship looked very like another, but one of Nelson's 'band of brothers' would spot the difference between a French ship and his own with no hesitation. The Royal Navy, as the largest organisation in the country after government itself, reaped the benefits of the Industrial Revolution. Standard weights and measures led to predictable performance of gunpowder, guns were safer and the dockyards could handle a massive workload of repair and outfitting in wartime. Before the American War of Independence critics always cited the impressive appearance of French dockyards, but during that long war the Royal Dockyards met all the challenges, whereas the French arsenals failed to keep pace with the demands of the seagoing fleet. It is reminiscent of the Duke of Wellington's comment on French logistics during the Peninsular War: 'The French system is like a magnificent set of harness, but if it breaks it requires the services of a saddler, whereas mine is an old harness held together by string, and if it breaks, I tie a knot and continue'.

This must not be taken as a slur on French ingenuity. The world's first ironclad was appropriately named *la Gloire*, and reflected great credit on her designer Dupuy de Lôme. It was, however, the Royal Navy's riposte, the huge ironclad 'frigates' *Black Prince* and *Warrior*, which totally eclipsed *la Gloire* and ended French boasting about stabling their horses in Westminster Abbey and staging a victory march to Buckingham Palace. Within a decade Britain's industrial might had wiped out the French lead. The British policy of allowing rivals to make the first mistakes, and then relying on its shipyards to build more, faster and better, always allowed the Royal Navy to be selective about new technology. The modern US Navy has pursued a similar course, not least because a huge fleet of ships cannot remain efficient if bombarded by constant innovations. Sadly, this was never understood by the French, who kept trying to find the 'silver bullet'; they went from the shell gun in the 1820s to the steam torpedo boat and onwards to the end of the century, producing clever ideas but never achieving their aim of blunting Britannia's trident.

It also pays to think hard about mass-production in an emergency. War mobilisations always trigger spending sprees, but within a few years the spectre of 'block obsolescence' appears. A good design can be updated to meet new threats, but no navy has yet achieved a one-for-one replacement of its reserve of wartime construction. It proved to be the Achilles heel of Admiral Gorshkov's great navy, which was matched by the rapid advances of Western technology to the point where wholesale replacements were too costly to contemplate. Gorshkov was caught in the same trap as Tirpitz; the master-plan depended on the enemy's meek acceptance of the state of affairs. By 1914 the cost of Germany's High Seas Fleet was causing severe distortions in the overall balance of defence expenditure. The lesson is, never assume that your enemy will react in the way that most suits you. In the words of a respected naval historian, 'Germany tried to play to a different set of diplomatic rules to outsmart the British, but the British responded by writing a third set of rules.'

Technical competence is a more complex subject altogether. Front-line navies of the 21st centuries do not suffer disasters because of the incompetence of their designers. The same was not true at the beginning of the age of iron and steam, when the switch from wood to iron introduced a new mathematical element. In the days of sail a three-decker's stability was not a problem. If the ship heeled violently in gale, she was likely to lose her rigging, sails and even masts before she reached a critical point of capsize. The judgement of when to reduce sail lay, therefore, with the captain, whose seamanship skills were critical. In practice wooden-hulled sailing warships caught fire, ran aground or were forced to surrender because there were no longer enough men to fight the ship. It was quite feasible for seagoing officers to make improvements to rig and to experiment with armament. Indeed, the Royal Navy encouraged captains to think for themselves. As late as the middle of the 19th century HM Ships' logs left space for comments on sailing qualities and recommendations for improvements.

The trouble came when iron, steam, large guns and armour came together, and senior officers could no longer be allowed a free hand in designing ships. It was now an engineering, hydrodynamics and physics-led science, and delicate calculations of stability, horsepower curves, metacentric height, angles of heel and so on could only be entrusted to naval architects. The loss of HMS *Captain* in 1870 was the last time a major warship was designed to conform to the ideas of a sea officer, in this case one of the gunnery branch. It has proved far better to get constructors to sea from time to time than to turn non-technical officers into amateur designers. There is also the sticky problem of who accepts responsibility when things go wrong, and one of the lesser-known requirements of an official design bureau is to be the final design authority. As a former Director-General Ships told me, 'You can devolve as much of the process as you like to industry, but somebody must carry the can if the bloody thing sinks.'

The delicate subject of sub-standard construction must also be looked at. The aim of commercial shipbuilders is to make as big a profit as they can, whereas the national navy wants the best value for its money. Major navies have over the years codified requirements for firefighting, stowage of shells and propellant and many lesser subjects. The purpose of the Principal Naval Overseer (PNO) or his equivalent is to ensure that all detailed features conform to official standards. Similar oversight must be maintained for firms manufacturing guns, ammunition, and engines. The formal acceptance of a new warship by a navy is much more than an occasion for celebration; it marks the moment at which full contractual responsibility is transferred from supplier to customer.

There has always been a conflict between official design bureaux and the officers who serve in their ships. Comparisons are drawn between the 'unimaginative' official team and the 'creativity' of commercial designs.

Official designers are working for a single customer, their parent navy, and have a limited choice: either to build the ship that the admirals say they want, or to persuade them that their requests are unreasonable. The Japanese resolved the dilemma by telling the constructors to do what they were told, and to leave the job of concealing any violation of relevant disarmament treaties to their superiors. Another common criticism is that modern ships are 'too comfortable', presumably to aid recruiting. The truth is that electronic systems require a comparatively small volume, and automation has reduced crew-numbers; the extra space is well used if devoted to providing sailors with air-conditioned mess decks. In fact, modern warships are volume-critical, not weight-critical in the way that warships designed and built before 1945 were.

Although the Royal Navy of the Edwardian period had no disarmament treaties to worry about, for most of the time design criteria were driven by the volcanic energy of the First Sea Lord, Admiral Sir John Fisher. The problem was that Fisher's imagination was not matched by any deep grasp of the fundamentals of design. He demanded ever-higher speeds, and was even prepared to 'leak' exaggerated speeds to the press. Many scholars have fallen into the trap of assuming that Fisher's ideas were all feasible, and his modesty never prevented him from claiming all the credit. The battleship *Dreadnought* was not Fisher's 'creation', but a logical progression from the previous design, as proposed to him by an experienced constructor, and following a trend towards a uniform armament of heavy-calibre guns at the expense of secondary armament. Fisher deserves the credit for persuading the government of the day to approve the extra cost, and for encouraging Portsmouth Dockyard during the construction period. In fact

Portsmouth Dockyard already enjoyed the reputation of the fastest building-times for battleships, and corners were cut by diverting four gun turrets intended for the previous class. The claim of 12 months from keel-laying to sea trials was, however, pure Fisher propaganda; the 'completion' date referred to first basin trials, and the ship took another two months to get to sea.

The propulsion of warships is another minefield for the non-specialist. Very few people serving in an average warship, apart from the engineers, have any idea of the relationship of installed power to speed, even less of the influence of extra weight on speed. Ever since steam was introduced, speeds have been quoted in reference books without any of these questions:

- Is the speed an average of a number of runs with and against the tide?
- What was the displacement of the ship during the trials (light, normal or deep load)?
- How long can the ship steam at high speed?
- What was the maximum speed achieved in service?
- Was the trial conducted by the shipyard or the navy?

Former constructor David K Brown once wrote an article on the number of ways in which sea trials can be rigged, citing examples that he had encountered. The early steam torpedo boats and destroyers ran preliminary trials under the supervision of the builders, often using specially trained stokers, keeping the displacement as light as possible and even in some cases using polished coal. In the 1890s the Admiralty invited a large number of shipyards to build the large number of destroyers needed to deal with the 'menace' of the French torpedo boats. The Controller's Department was taken aback when a small Scottish shipbuilder, Hannah Donald, wrote a letter to ask what the ideal depth of water was for increasing the speed. On investigation it turned out that the two leading yards, Thornycroft and Yarrow, then sited on the River Thames, were running trials over a measured mile off the Maplin Sands. The ground wave caused by the destroyer's movement through the shallow water created a 'hump' which added as much as a knot to the speed. The Admiralty had been greatly puzzled by the failure of most destroyers to equal their trials speeds in service; it reacted swiftly by insisting that future trials be run on a deepwater measured mile.

Maintaining a good design team has always entailed the recruitment of well qualified people. In peacetime staffing is relatively easy, but the pressures of war and preparations for war can inflict severe strain. A major navy starts to generate large numbers of specialised designs as the nation drifts towards war, and mistakes (though not many) are often made under stress. How else to explain the fact that Bureau of Ordnance (BuOrd) team working on the Iowa-class battleships in 1940 could design the barbettes for the triple 16-inch guns with 18 inches less diameter than needed. The plan had been to use 16-inch guns retained from capital ships scrapped at the time of the Washington Treaty, and although the Bureau of Ships (BuShips) was less than half a mile away, apparently there was minimal contact between them. The solution was to design a new gun with less recoil, so that its turret could fit into the smaller barbette. Probably the United States was the only country in the world capable of achieving such a feat, but it was an expensive solution to a major administrative blunder.

The final heading of my list of influences is the operational environment. The area in which a ship is likely to spend most of its time may be relatively calm or stormy. The wave period and the steepness of the waves, as well as wind speeds, determine the behaviour of ships and their ability to fight. Habitability in a ship operating in an area for which she was not designed can seriously degrade the efficiency of the crew, and in extreme cases affect weapon performance.

It was one of the boasts of the Victorian Royal Navy that it lost so few ships from 'stress of weather'. The sea is always unforgiving, and even the relatively calm Mediterranean can be stormy, so stormy that an Italian destroyer was lost in a storm in 1942. On 8 September 1923 seven US Navy destroyers ran aground off Point Arguello in California, becoming total losses. During the Second World War, on 18 December 1944 units of the Pacific Fleet ran into a typhoon and three modern destroyers capsized.

Modern research suggests that prolonged seasickness is the cause of accidents on board (handling machinery, for example) and for inefficiency in operating complex electronic systems. It is signifcant that long-hulled escorts in the Battle of the Atlantic drew fewer criticisms for unpleasant rolling and pitching than fine-hulled destroyers and short-hulled corvettes and trawlers. As a result modern warships make more use of active fin stabilisers to mitigate the worst effects of weather. During the Second World War the US Navy showed that self-compensating fuel tanks allowed its ships to operate at constant trim.

For all these reasons I wanted to write this book. After writing, lecturing and enjoying this multi-layered subject for over three decades I am still surprised by its complexity. I have been very privileged to have had so many chances to talk to designers, to visit the shipyards where their creations take shape, and finally to go to sea in many of them. It has also given me opportunities to meet some of the most interesting range of people, from admirals and senior naval architects to survivors of calamities such as the sinking of HMS *Sheffield* in 1982. I have also tried to see ships as the 'total systems' that they are, even to the point of visiting the factory at Bourges in France where the Exocet missile is manufactured.

CIVIL WAR MONITORS

US NAVY 1861–1937

The shallow-draught monitor was the US Navy's great contribution to naval warfare, intended to meet the peculiar conditions of the US Civil War. On the East Coast, the huge Navy Yard at Norfolk, Virginia, had been evacuated shortly after the bombardment of Fort Sumter in April 1861 and passed into the hands of the Commonwealth of Virginia and, latterly, the newly-created Confederate States of America. The Union Navy needed time to mobilise its industrial strength, and resorted to the proven weapon of blockade to prevent the Confederacy from exporting its cotton crop to pay for the war.

The Union Navy was lucky to have been pestered by the Swedish-born inventor John Ericsson, who had ideas about armouring a flat, low-freeboard deck and a revolving turret carrying the heaviest guns available. His ideas suddenly became fashionable when Union spies reported that the Confederate States Navy was building an 'ironclad' capable of driving the Union blockaders from the mouth of the Chesapeake Bay. The Union Navy was particularly lucky in Ericsson because he was a maritime engineer who had played a major part in the development of the screw propeller, was a competent engineer, and had designed several ships. He was therefore no amateur enthusiast playing a hunch, that bane of navy administrators.

As soon as the Union Navy's senior officers and civilians had evaluated Ericsson's design for a turret ironclad to counter the Confederacy's ironclad floating battery an order was placed. His proposals were radical: a flat-bottomed hull with bilges angled at 35 degrees to the horizontal, drawing only 10ft 5in when ready to go to sea. The engines, also designed by Ericsson, had two pistons in a single cylinder operating 'driving levers' connected to the propeller-shaft. Speed was a modest 6kts, and the freeboard of the hull was only 18in.

By far the most radical feature was her armament of two 11in Dahlgren smooth-bore shell guns, which were mounted in a cylindrical revolving turret. As the turret was heavily armoured, independent auxiliary steam engines were needed to rotate it at 2.5 turns per minute, as well as running ventilation fans. Some idea of the weight penalty incurred can be gained by looking at the armouring of the turret: eight layers of 1in iron were bent around the 20ft diameter turret. The upper or weather deck consisted of horizontal oak beams covered with two half-inch plates, while the sides were protected by 5in iron armour over 25in of oak. Apart from the turret, the only other deck structure was a small pilot house forward. The low freeboard and flat weather deck were intended to make smooth-bore spherical shells bounce off, whereas the turret was intended to survive direct battering.

The little ironclad, in reality no more than an armoured gun-platform suitable for inshore operations, was laid down in October 1861. She was named *Monitor*, not to commemorate the

The USS Monitor in the James River in July 1862. The Ericsson turret shows dents made by Confederate artillery.

large aquatic lizard, as some have tried to claim, but to suggest an alert sentry. She was commissioned on 25 February 1862, only 12 days before the CSS *Virginia* emerged from her lair in the James River.

The Dahlgren guns, with their familiar 'soda water bottle' shape, were as their name suggests, designed by Admiral John A Dahlgren, the pre-war US Navy's Chief of the Bureau of Ordnance. He had undertaken to reform the parlous state of US naval ordnance in the 1840s, and delivered his first 9in smooth-bore to Washington Navy Yard in May 1850. Next came an 11in gun in 1851. The standard powder charge was 15lbs, but in battle it proved unable to penetrate the armour of the CSS *Virginia*, better-known to history by her original US Navy name, the *Merrimack*. Dahlgren thereafter increased the charge to 20lbs, and even 25lbs, with no ill-effects. The story that the *Monitor* fired her 11in Dahlgrens with reduced charges is incorrect; she was firing the standard 15lb charge.

The *Monitor*'s moment of fame came early in her short life. On 8 March the *Virginia* sortied to attack the Union fleet in Hampton Roads, sinking the sailing frigate *Congress* and driving the *Cumberland* ashore, before vanishing upriver. Something very like a panic seized the Union Navy's senior echelons, to say nothing of the blockading squadron, but salvation was close. That night the little turret ship arrived, and was ready for battle. The *Virginia* appeared next morning, confident of another easy victory.

The Battle of Hampton Roads which followed was a remarkable affair and, as the first battle between ironclads, transformed the attitudes of all navies on the subject of ironclads and the revolving turret. For three hours the two ships manoeuvred and exchanged fire, but did no serious damage to each other. Even the temporary blinding of Lieutenant Worden, commanding officer of the *Monitor*, by a shell bursting outside the pilothouse, had no effect on the outcome. The *Virginia*, unable to make any impression on the 'cheesebox on a raft', and cheated of an opportunity to ram her tiny opponent, went back upriver and did not emerge again. Although the blockading squadron was saved, the existence of the big Confederate ironclad helped delay the Union Army's advance on the Confederate capital Richmond during the Peninsula Campaign. Her menacing presence paralysed Union efforts for another two months, and she was not neutralised until 11 May 1862, when the Confederates scuttled their remarkable creation.

It is often claimed that the Battle of Hampton Roads forced the Royal Navy to take the turret seriously, but this is not true. In reality the Admiralty had already tested the Cowper Coles turret, and had decided to build two competing turret-ship concepts, the purpose-built *Prince Albert* versus the *Royal Sovereign*, converted from a wooden-hulled ship of the line. The Secretary of the Admiralty noted only that the 'recent events off the coast of Virginia' eliminated the need to justify the adoption of Captain Cowper Coles's turret to the press and Parliament.

USS *Monitor*

Laid down 25 October 1861, launched 30 January 1862, commissioned 25 February 1862, built by Continental Iron Works, Greenpoint, New York

Displacement:	987 tons
Dimensions:	172ft x 41ft x 10ft 5in
Machinery:	1-shaft Ericsson reciprocating steam, 320ihp
	2 Martin boilers
Speed:	6kts
Armament:	2-11in smoothbore ML (1 x 2)
Armour:	4.5in–2in side; 9in–8in; 1in deck
Coal:	100 tons
Complement:	49

The revolutionary little 'rivergoing raft' was lost on 31 December 1862, when she foundered while in tow. Information gleaned by maritime archaeologists from examination of the wreck in recent years suggests that the cause of her loss was separation of the upper part of the hull from the lower. Whatever merits she displayed, a seagoing capability was not one of them.

The limited success of the *Monitor* in frustrating the *Virginia*'s primary mission led inevitably to a state of public euphoria, and profoundly affected the nature of the sea war. What the distinguished Union Navy historian Donald J Cannery has called 'monitor fever' took over the North, and the monitor-type was seen as a panacea for all the Union Navy's problems. In practical terms, this meant that monitors were now seen as capable of attacking fortifications as well as Confederate ironclads. In all 60 monitors were started during the war, 37 of which were completed by the end of 1865. In contrast only two oceangoing ironclads were built.

In December 1862 the first of an improved monitor type, the *Passaic*, was completed, having been authorised less than a week after the Battle of Hampton Roads. Eight more followed from East Coast builders by April 1863, but the attempt to build monitors on the West Coast was only a partial success. The *Camanche* was not commissioned until May that year, having been shipped in her component parts by sea from the East Coast. There were several major changes to rectify the more glaring faults of the *Monitor*: a more ship-like hull form with a slight sheer, moving the pilot house to the top of the turret, more armour and a more powerful armament. The intention was to arm them with two 15in Dahlgren guns, but these were in short supply, so the wartime armament of most was one 11in and one 15in; the 11in in the *Lehigh* and *Patapsco* was a Parrott rifled gun.

Seven of the *Passaic*s took part in the unsuccessful attack on Charleston, South Carolina, on 7 April 1862, and their shortcomings were soon obvious. They proved vulnerable to hits at the base of the turret, which tended to jam it; the pilot house offered very little protection to personnel, and the rate of fire of the 15in gun proved painfully slow. Five to seven minutes per round was all that could be achieved. Used properly, against enemy ironclads rather than fortifications, they were successful; when the *Weehawken* encountered the ironclad CSS *Atlanta* in 1863 her 15in shells penetrated 4in of iron and 18in of wood, and forced the *Atlanta* to surrender after only five hits. On the other hand, the same monitor foundered on 6 December 1863 when a storm caused water to enter the ship through her hawsepipe. With no internal subdivision these ships had little reserve of buoyancy; the *Tecumseh* and the *Patapsco* sank quickly after setting off Confederate mines during the attack on Mobile, Alabama. In contrast, when proper precautions were taken, they could survive bad weather; the *Lehigh* made port after encountering a Force 10 gale off Cape Hatteras and having her deck 4ft under water.

Passaic class

Passaic, Patapsco, Nahant, Montauk, Sangamon, Weehawken, Nantucket, Catskill, Lehigh, Camanche.

Displacement:	1335 tons
Dimensions:	200ft x 45ft x 11ft 6in
Machinery:	single-shaft reciprocating steam
Speed:	7 kts
Armament:	1-11in SB, 1-15in SB
Coal:	150 tons
Complement:	75

Other monitors followed, the one-off *Onondaga*, the double-turreted *Miantonomoh* class, the converted frigate *Roanoake* (the sister of the *Merrimack*, given three twin turrets), the *Dictator* and *Puritan*, the *Canonicus* class, the *Kalamazoo* class and the shallow-draught *Milwaukee* class, 24 ships in all. But the strangest were the 20 *Casco* class, all ordered in 1863, although only nine were commissioned in 1864–65.

Ericsson cannot be blamed for the gross miscalculations of weights in this class as they were designed by Alan Stimers. They were intended to operate in the shallowest possible waters; the designed draught was 6ft 4.5in and 15in of freeboard, and ballast tanks were provided to reduce the freeboard still further when going into action. To everyone's amazement the second unit of the class, the *Chimo*, without the twin 11in turret or stores, floated with only 3in of freeboard. So unsuited were they for their intended role that a drastic decision was made to complete five of them as spar torpedo boats, with thinner decks, Wood-Lay spar torpedoes, no turrets, and coal bunkerage reduced by 70 tons.

Although designed for a speed of 9kts, the *Casco* class never exceeded 5kts, making them useless for attacking with spar torpedoes. The *Casco* was used to clear mines from the James River in 1865, but that was probably their only achievement of any note. Improvements to armament were made, but at the cost of deepening their hulls by 22in. All were sold for scrapping in 1874–75.

The US Navy remained fixated by the alleged virtues of the big monitors, to the extent of perpetrating one of the great administrative frauds in naval history. The *Puritan* and the four *Miantonomoh* class were given 'rebuilds' in the 1880s, but this was a device to get around the refusal of Congress to vote funds for new construction, and in the words of one US historian, 'the nameplate was unscrewed and used for a new ship'. The 'rebuilt' *Miantonomoh* class became the *Amphitrite* class, with two names repeated, the *Monadnock* and *Miantonomoh*, but the original hulls had been scrapped some years earlier.

When eventually funds became available in the late 1880s five more monitors were built, the *Monterey* (BM-1) and the four *Arkansas* class (BM-7 to BM-10), all armed with a twin 12in armament in a single turret. They were highly regarded at the time, but the concept was obsolete. After acting in their designed coast defence role during the war against Spain in 1898 they were reduced to subsidiary duties. Their low freeboard made them ideal as submarine tenders, and as a result they survived until the 1920s; the longest-lived was the *Cheyenne* (ex-*Wyoming*), which was decommissioned in 1926 but not sold until 1939. Names were changed to allow the old names to be used for new battleships.

Surprisingly, some of the *Passaic* class survived until the Spanish-American War in 1898, but thereafter they were stricken in quick succession. The *Catskill* was the only one to spend much time in commission. The only other country which built to the classic monitor design was Sweden; the so-called monitors built by the Royal Navy in the First World War were in fact highly specialised shore-bombardment vessels.

Conclusion

The 'monitor mania' which ensued after the Battle of Hampton Roads is an example of what can happen when a new type of warship scores what is essentially a 'draw' or a minor victory. With a war on, the type's capabilities are not evaluated properly, and a knee-jerk reaction is to build more of the same. The Union Navy also made an error in trying to use monitors for missions outside their capabilities. Against a stronger opposition with better industrial resources more than four might have been lost (two foundered and two mined).

The reaction to the *Monitor*'s performance at Hampton Roads can be forgiven; the Union Navy was severely shaken by the threat to its largely wooden fleet. To everyone, therefore, the *Monitor* seemed heaven-sent, but the resources frittered away on such a large force of successors, some of dubious value, could surely have been spent on more effective vessels.

The verdict on their seakeeping is not totally negative. As already said, the *Lehigh* survived a Force-10 gale, while the *Monadnock* braved Cape Horn in 1865. But the commanding officer of the *Miantonomoh* was surely exaggerating when he said after crossing the Atlantic in 1866 that he could 'whip any ship in the Royal Navy'. Faith in one's command is all very well, but the boast merely confirms that the reputation of the monitors was inflated. Not until the 1890s could the US Navy really be said to have kicked the monitor habit.

TURRET SHIP HMS *CAPTAIN*

ROYAL NAVY 1866–1870

The loss of the new, technically advanced ironclad HMS *Captain* remains one of the worst peacetime disasters in history. To understand the chain of events leading up to the tragedy we must retrace our steps to the earlier Russian War of 1854–56 (popularly but incorrectly known as the Crimean War). In 1855 a talented British gunnery officer, Captain Cowper Phipps Coles, had designed and built a 'cupola' on a raft improvised from casks and planking. Named the *Lady Nancy*, the gun raft could be rotated to give all-round fire, She proved a great success in the bombardment of Taganrog in the Sea of Azov, and Coles patented his invention four years later.

Within six months the Admiralty ordered a prototype turret, which was installed in the floating battery *Trusty* in September 1861. The trials were so successful that the Admiralty ordered an iron-hulled coast defence ship armed with four turrets, the *Prince Albert*, in February 1862. Two months later a conversion from a wooden ship of the line, the *Royal Sovereign*, was ordered to provide a comparison.

The Admiralty was impressd by the performance of the two coast-defence turret ships and by the capabilities of the USS *Monitor*, but recognised that an oceangoing warship was needed to meet the Royal Navy's worldwide commitments. But the technology available created a number of major problems. Fuel consumption was still too heavy to permit a battleship to cross the Atlantic under steam alone, and this dictated the retention of masts and rigging. The Chief Constructor of the Navy, Sir Edward Reed, insisted that the upper deck of a fully rigged ship was not a suitable site for a turret. Coles chose to ignore such objections, proposing in 1859 a rigged ship with *ten* turrets, a project which impressed credulous commentators but merely evoked scorn from the Chief Constructor's staff.

Both Reed and Coles were opinionated men, but Reed enjoyed the prestige of being the country's most distinguished naval architect, whereas Coles was so obsessed with his 'big idea' that he refused to be deflected by such side issues as physics and hydrodynamics. In this he was one of a long line of promoters who believed they knew more about warship-design than the professionals. What is worse, they lobby the media and politicians, two groups not known for their insights into high technology. Naval officers are particularly fond of drawing up 'paper designs' and then accusing qualified designers of being reactionaries for raising practical objections.

Coles was given permission to prepare a rigged two-turret ship, with the help of a constructor, but work was stopped to allow the *Royal Sovereign* to be evaluated. He tried again in 1864 with a single turret design, and once again the Admiralty provided technical help in the form of the Chief Draughtsman at Portsmouth Dockyard. Early the following year the Admiralty set up a committee to 'obtain the unbiased opinion of practical naval officers'. Coles demanded the

HMS Captain *was the product of one man's enthusiasm and political influence.*

right to nominate half the members of the committee, but not surprisingly, this was turned down. He was invited to attend, or alternately, when it was learned that he was ill, to send a representative, but he declined both offers and refused even to submit questions to the committee.

The committee concluded that the turret must be adopted, but did not favour Coles's single turret ship because of its poor arcs of fire. What was recommended was a two-turret design with two 12-ton (9in) guns in each, or a single 22-ton (12in) gun, and the Admiralty decided to build a ship which broadly met the committee's recommendations. This was HMS *Monarch*, armed with two turrets containing a pair of 25-ton (12in) rifled muzzleloaders each, and fully rigged. Reed did not like the idea of combining turrets with a full ship-rig, although he added some features intended to alleviate the problems.

Coles was not at all happy with the design of the *Monarch*, and lobbied for a design incorporating his own ideas of low freeboard. His views were expressed vociferously in public, and amounted to venomous personal attacks on the Controller, Sir Spencer Robinson, and his department. Coles' consultancy was cancelled in January 1866 because of these attacks, but when he climbed down, saying he had been misunderstood, he was reinstated. The Board was by now fed up with Coles and told him on 24 April that the features of the *Monarch* to which he took exception were there at the request of the Board, and there was no point in further discussions. The aggrieved inventor now turned to his friends in the press and Parliament, demanding the right to build a turret ship to his own design.

As a result of intensive lobbying the Admiralty was forced to agree to fund a second turret ship, its building supervised by Coles. On 8 May 1866 he told the Admiralty that he had selected Laird Brothers' Merseyside yard, successful builders of iron warships. By mid-July Lairds submitted two designs, one with twin shafts and one with a single shaft. Six days later Reed commented that the design seemed 'well considered and well contrived', provided '*we take for granted that the deck is high enough*'. He accepted a freeboard of 8ft, but Spencer Robinson thought it was too low. The First Lord of the Admiralty, Sir John Pakington, however, told Coles that he approved the building of the ship, on the understanding that responsibility was shared *entirely* between him and Lairds.

The formal contract prepared in November 1866 took account of Coles's illness by substituting the words 'Controller of the Navy' for Coles's name throughout. Reed, on the other hand, regarded Lairds as having full responsibility, and told his staff to mark all drawings as 'No objection is seen', rather than the normal 'Approved'. But the design of a warship does not lend itself to such split responsibility, and Spencer Robinson should have seen the danger. (Reed should also have admitted the risks inherent in building a low-freeboard, fully rigged turret ship.)

Lairds seem to have assumed an adequate margin of stability, but in fact the height of the centre of gravity of the *Captain* in load condition rose from an estimated 21ft 6in in 1866 to a calculated 22ft 3in in 1870, and an inclining experiment verified this latter figure. The builders chose to ignore Reed's misgivings, for which they must be blamed. Reed's assistant Barnaby also reported that excessive weights, up to 860lbs, were being taken on board. Modern experience suggests that this discrepancy may have been caused by an error in the original calculations, but either way, it reflects on the design abilities of the builders.

After launch the ship's condition and draught of water were checked, and she was found to be 735 tons overweight and floating 22in deeper than predicted. Her freeboard turned out to be only 6ft 7in, not the 8ft planned, but still no alarm bells rang. She was a novel ship, and her characteristics were not yet fully understood.

HMS *Captain* was completed in March 1870 and immediately won high praise from Coles's supporters for her performance during competitive trials against her rival HMS *Monarch*, and on her first commission in the Channel Fleet. When she returned to Portsmouth in July that year, at the end of her second commission she seemed to justify all her admirers' hopes, but one of Reed's staff then conducted an inclining experiment. This showed that the righting moment—the point beyond which the ship was bound to capsize—was 20 degrees. The righting moment fell away as soon as the deck-edge became immersed.

The fleet was exercising in heavy weather off Cape Finisterre on the night of 6 September 1870. Admiral Sir Archibald Milne, the Commander-in-Chief of the Mediterranean Fleet, had visited the *Captain* that afternoon, and saw that the sea was washing over the lee of the upper deck, and that the deck-edge was under water at 14 degrees. When the wind speed rose during the evening, orders were given to take in sail, and at midnight, when a fierce squall blew up, Captain Burgoyne ordered the topsail halyards to be cut. Before this could be done the *Captain* heeled over and sank, taking with her all but 18 of her crew, out of a total of 490, including Captain Cowper Coles and Captain Burgoyne.

Burgoyne had come on deck and took charge of the frantic efforts to reduce the effect of the wind, having reached the bridge 'scantily dressed' in spite of the rain. He gave a rapid succession of orders, one of which was to tell the officer of the watch to check the angle of heel on the ship's clinometer. This was showing 18 degrees, not an alarming heel in other big ships, and it suggests that Burgoyne had begun to doubt in the stability of his ship. The senior midshipman had mustered the relieving watch on the forecastle, when the squall struck the fleet. The

200 hands on deck were all thrown into the sea, while the 300 men below decks were nearly all overwhelmed in their hammocks. The fate of the stokers must have been even worse, as the *Captain* lay for a few moments on her beam ends and then rolled bottom up and disappeared. Survivors reported hearing agonised shouts and screams through the ventilators and hatchways.

To make matters worse, the big frigate HMS *Inconstant* passed within 50 yards of the spot only ten minutes later. The total darkness and the noise of the wind, however, prevented her lookouts from seeing anything, and the survivors struggling in the water watched her lights disappear. The only boats to float free were a 37ft steam pinnace and a 36ft oared pinnace, floating upside down. Seven survivors, including Captain Burgoyne and James May, a gunner, clung to the upturned keel. When the other pinnace came near May jumped across, and the men on board tried to rescue Burgoyne with a boathook. But when it slipped from his grasp they offered him an oar, to which he replied, 'Save your oars boys, you'll need them', and slipped away into the darkness. May was one of only 18 men to survive the loss of HMS *Captain*, and they made a successful landfall at Corcubion Bay on the coast of Spain.

In accordance with naval practice the survivors were court-martialled and duly acquitted, but the Court took it as its duty to 'record the conviction they entertain that the *Captain* was built in deference to public opinion expressed in Parliament and other channels…'. The First Lord at that time, Hugh Childers, had been one of Coles's vociferous supporters while his party was in Opposition, and had demonstrated his faith in Coles's ideas by sending his only son to sea in *Captain*. He issued a paper justifying his actions and attacking everyone else; he added to this unsavoury chapter by forcing the resignation of Spencer Robinson, one of the more talented officers in the Royal Navy. In his place Childers tried to appoint one of the Laird brothers, one of the people most responsible for the errors in the calculations! Some of Coles's more fanatic adherents claimed that the ship had been lost through defective seamanship.

Edward Reed resigned before the loss of the *Captain*, worn down by constant sniping and lack of support from his political masters. Spencer Robinson was hardly exaggerating when he called the resignation a national disaster. Reed had instigated and directed major improvements in every aspect of design procedure, and also had experience of the commercial world. After his resignation he designed a number of ironclads, but these lacked the inspiration of his work for the Admiralty. He entered Parliament, and launched a long series of vituperative attacks on his successors. There is little doubt that his intemperate attacks alienated many of his original supporters, but he was honoured by being elected Vice President of the Institute of Naval Architects in 1865. If vindication were needed, HMS *Monarch* proved successful, and although soon outdated, remained operational for nearly 30 years. Unlike many Victorian battleships she saw action, taking part in the Bombardment of Alexandria in 1882.

HMS *Captain*

Laid down 30 Jan 1867, launched 27 March 1869, completed Mar 1870, built by Laird Bros, Birkenhead

Displacement:	7767 tons (load)
Dimensions:	320ft (pp) x 53ft 3in x 24ft 10in
Machinery:	2-shaft 4-cyl reciprocating, 5,400ihp; 8 boilers
Speed:	15.25kts
Armour:	8-4in belt, 10-9in turrets, 7in CT
Armament:	4-12in (25t) RML (2 x 2), 2-7in (2 x 1)
Complement:	500

Conclusion

Looking at the evidence over 130 years later, it is clear that four major errors led to the tragic loss of HMS *Captain*.

The first was the decision to build a fully-rigged, low-freeboard turret ship. It was quite simply wrong, and both Spencer Robinson and Reed had warned of the impracticalities on more than one occasion. The bad-tempered and almost paranoid attitude of Captain Coles to any questions misled the minds of the public, the press and ministers, who chose to see the differences as some sort of David v. Goliath struggle of the plucky inventor against the bureaucrats. In such an atmosphere it was impossible to hold a reasoned debate.

Second was the failure to define responsibility between Laird and the Controller. This was exacerbated by Coles's illness, which prevented him from playing any part in the detailed design. As David K Brown has commented elsewhere, Lairds were very quick to wash their hands of responsibility after the loss, but there can be little doubt who would have claimed the credit if HMS *Captain* had been a success.

The third major error was Laird's failure to pay any attention to Reed's clearly expressed warning that the ship's centre of gravity was higher than they thought. In part this can probably be blamed on Childers's partisanship, which gave Coles and the builders the dangerous idea that they needed to pay no attention to the Controller or the Chief Constructor of the Navy.

The fourth error was Laird's incorrect calculation of the weights in the ship, which caused her to float deeper in the water than planned. Taken with the other three errors, this failure sealed the fate of the *Captain*, and her proud name never again appeared on the Navy List.

Vitse Admiral Popov and *Novgorod* Coast Defence Ships

Russia 1872–1912

In the 1860s a number of authorities proposed maximising the beam of battleships to the point where it rivalled length. The aim was to shorten the hull and thereby reduce the area protected by armour, but it also improved manoeuvrability. The distinguished British naval architect Sir Edward Reed advocated such solutions, and Russian Vice Admiral A A Popov took the idea to its logical extreme, a flat-bottomed circular hull.

As always, there was also a political dimension. After the crushing Anglo-French victory in the Russian War, the Treaty of Paris banned Russia from building and maintaining a battle fleet in the Black Sea. The Russian Admiralty feared another British incursion into the Black Sea, mindful of the destructive campaign in the Sea of Azov in 1855 and the massive bombardment which destroyed the Kinburn fortress. What the Russians wanted was a force of armoured and well-armed shallow-draught coast defence ships to guard the Kerch Straits and the mouth of the Dniepr River.

In 1870, after tank tests, Popov had built a small 24ft circular steamer to test his ideas and persuaded the Imperial Russian Navy to adopt his ideas for a fleet of ten circular ironclads. They were to serve as 'floating forts', capable of all-round fire, armed with the largest possible guns and well protected by armour. The plan proved too expensive, however, and only two were built, the *Novgorod* and the slightly larger *Vitse Admiral Popov* (*Rear-Admiral Popov*), originally laid down as the *Kiev* but renamed in honour of her designer. The former was commissioned in 1873 and her sister three years later.

The design embodied unusual features. In the *Novgorod* the side armour extended from the deck edge 1ft 6in above the waterline to 4ft 6in below it. It was 9in thick for the upper 3ft with a 7in lower strake. The hull was stiffened internally by heavy iron stringers. The 2.75in thick deck was 5ft 3in above the waterline. The barbette was protected by 9in armour and extended 7ft above the deck, and the funnels had 4.5in armour to a height of 3ft. The hull had heavy 27in wooden sheathing outside the armour and was coppered. The ship was armed with a pair of 11in rifled breech-loading guns on retractable mountings in a fixed barbette (firing over the top). According to some sources, during a later modernisation these guns were replaced by 8in breech-loaders.

The *Vitse Admiral Popov* was similar in most respects, but the side and barbette armour had

The Popovki's circular hull form theoretically might offer improved manoeuvrability and all-round firepower.

7in plates added, though separated from the main hull by a packing of 4.5in of wood. The 11in/20 cal guns in the *Novgorod* gave way to 12in/20 cal in her half-sister, the deck armour was increased to 3in, and she was given more powerful engines. The light armament of the two ships differed slightly.

The *Novgorod* was driven by 6-shaft reciprocating engines, with steam generated by eight cylindrical boilers. Output was 3000ihp, good for a theoretical maximum speed of 7kts. The *Popov* was given machinery developing about 4500ihp, for a theoretical maximum speed of 8.5kts. Neither achieved their designed speed. Later the outer shafts were removed from both ships, and two boilers. Coal capacity was 160 tons in the *Novgorod*, and 170 tons in her half-sister.

The *Novgorod* was built in sections at St Petersburg and then reassembled at the Nikolaev naval shipyard near Odessa. The *Vitse Admiral Popov* was also built at Nikolaev, but from the keel up. In service the Russians nicknamed them *Popovki*, a name mutated into 'Popovkas', by which they became widely known to naval history students and enthusiasts.

The comparatively light hull allowed a heavy scale of protection, 20 per cent of the *Novgorod*'s displacement and 34 per cent of the *Popov*'. But in other respects they were a dismal failure. They were too slow to stem the current in the Dniepr, and proved very difficult to steer. In practice the discharge of even one gun caused them to turn out of control, and even contra-rotating some of six propellers was unable to keep the ship on the correct heading. Nor could they cope with the rough weather which is frequently encountered in the Black Sea (as seen in the great gale of 1854 which wrecked the British effort to supply the army besieging Sevastopol). They were prone to rapid rolling and pitching in anything more than a flat calm, and could not aim or load their guns under such circumstances.

Both served as coast defence ships for the best part of three decades, but at the beginning of the 20th century they were laid up at Sevastopol and served thereafter as stores ships. Both were scrapped just before the outbreak of the First World War in 1914.

NOVGOROD

Laid down 1872, launched 1873, completed 1874

Displacement:	2491 tons (normal), 2706 tons (full load)
Dimensions:	101ft x 101ft x 12ft 6in (maximum)
Machinery:	6-shaft reciprocating steam, 8 cylindrical boilers, 3000ihp (later reduced to 4-shaft 2000ihp)
Speed:	6kts
Armament:	2-11in/20 cal BL (1 x 2); 2-88mm (2 x 1); 2-2.5pdr; spar torpedoes
Armour:	7–9in side (wrought iron), 9in barbette
Coal:	160 tons
Complement:	8 officers, 120 ratings

VITSE ADMIRAL POPOV (EX-KIEV)

Laid down 1874, launched 1875, completed 1877

Displacement:	3550 tons (normal), 3990 tons (full load)
Dimensions:	120ft x 120ft x 13ft 6in (maximum)
Machinery:	6-shaft reciprocating steam, 8 cylindrical boilers, 4500ihp (later reduced to 4-shaft: 3066ihp)
Speed:	8kts designed, reduced to 6kts
Armament:	2-12in/20 cal BL (1 x 2), 8-88mm (reduced to 6), 2-1pdr revolving MGs
Armour:	16-14in wrought iron on sides, 16in barbette
Coal:	170 tons
Complement:	203 officers and ratings

CONCLUSION

The lesson of the Russian circular ironclads must surely be that no single design feature must be allowed to dominate at the expense of 'normal' requirements. In the case of the Novgorod class, the circular hullform produced entirely retrograde side-effects such as poor steering. The luxury of tank-testing was only at an early stage of development in the late 1860s, and we can be certain that even a moderate degree of care in supervising a tank-test would have revealed the lack of directional stability and the lack of propulsive power. It is significant that an 'unscientific' powered scale model first exhibited in 1974 displayed the same characteristics as the real 'Popovkas'. It could not be steered because the rudder had no effect at all, it was very slow, but it had no difficulty in making a 360 degree turn!

On a more scientific note, a circular hullform offers the most resistance and drag imaginable, and clearly the Russians recognised eventually that the outer propellers were contributing

little or nothing to propulsive efficiency. The excessively shallow draught and flat bottom did not help either, making the ships behave like skimming-dishes. The heaviest armament and protection are no use if they cannot be brought into action quickly and predictably, and in naval warfare that means the 'platform'.

The Glasgow shipbuilder John Elder had been a keen exponent of circular ship theory in the early 1860s, but when his shipyard won an order twenty years later to build a yacht for the Tsar of Russia he did not apply his theories so single-mindedly. The requirements included maximum longitudinal and transverse stability, as well as minimal motion in a seaway. Elder's designers examined the possibility of using a circular hullform to reconcile these requirements, but soon concluded that such a vessel would lack directional stability, i.e. be very difficult to steer. It would have made the Imperial Russian Navy a laughing stock if the Supreme Autocrat's yacht could not steam in a straight line while reviewing the Fleet. By then the dismal reputation of the 'Popovkas' was well-known, and the concept was clearly unsuitable. Elder therefore modified Popov's ideas, selecting a 'turbot shape' by providing a bow and stern of conventional type. The yacht, named *Livadia*, proved a great success and was noted for her comfort and good seakeeping.

In fairness to Popov, his circular coast defence ships were never intended to fight a conventional sea battle, and if the basic faults could somehow have been avoided, they would have been adequate for their very limited mission. But, as things turned out, the Imperial Russian Navy made the right decision when it cut the programme from ten ships to two.

A happy footnote to the story of the Popovkas is provided by the Imperial Yacht Livadia, *which had a conventional bow and stern, but a circular hull. She proved comfortable and a good seakeeper.*

Armoured Rams
HMS *Polyphemus* and
USS *Katahdin*

Royal Navy 1882–1903 and
US Navy 1896–1909

Although the tactic of ramming and sinking an enemy ship had been a feature of ancient naval warfare, it became attractive 2000 years later, in theory at least, when iron-hulled warships appeared. Here at last, it seemed, were ships with sufficient strength in their hulls and sufficiently powerful engines to sink an enemy ironclad. In part this was because the new ironclads seemed impossible to sink by gunfire. What the theorists ignored was the fact that wooden ships had also been extremely difficult to sink by gunfire. In Nelson's day it was possible to disable an opponent by killing enough of her crew to facilitate capture by boarding, or by disabling her masting and rigging. The losses in naval actions at sea were usually through grounding or fire.

The perceived value of ramming rested on the sinking of the Union Navy frigate USS *Cumberland* by the Confederate ironclad CSS *Virginia* in the Battle of Hampton Roads in 1862, and the Austrian victory over the Italians in the Battle of Lissa in 1866. Closer analysis of Hampton Roads might have shown that the *Cumberland* had been disabled by the *Virginia*'s gunfire before she was finished off. At Lissa, the Austrian Admiral Tegetthof ordered his ships to get to close quarters and ram because his ships had so few guns capable of penetrating Italian armour. His orders led to eight ramming incidents, of which only one led directly to the sinking of the broadside ironclad *Re d'Italia* by Tegetthof's flagship *Ferdinand Maximilian* after her steering had been disabled. The iron coast defence ship *Palestro* was badly damaged by ramming, but a shell started a fire in her poop, and she was destroyed by a magazine explosion.

We can now see that ramming tactics were a very clumsy way to sink enemy ships, and accidental losses were much more likely (ramming was frequently more fatal to the rammer than the rammee). In practice the damage was usually slight, and the sinking was caused by inadequate watertight subdivision and failure to close watertight doors. But the advocates of ramming chose to ignore these factors, and as a result more and more ships were built with strengthened bows and projecting spurs. They also insisted that ships should be capable of end-on fire to neutralise the enemy's fire during the final approach. The most obvious result of this obsession was the large number of what we now think of as freakish hybrids such as turret rams and central battery ships, all involving complex layouts of heavy guns.

The men who tried so hard to develop new tactics to make the best use of the early ironclads can be forgiven for promoting the idea of ramming, but their successors had less excuse. In the mid-1870s the Royal Navy's talented designer, Chief Naval Architect Nathaniel Barnaby and one of his constructors, W J Dunn, prepared an imaginative design for a fast ship with a cigar-shaped hull, an armament of up to five submerged torpedo-tubes, and as many as forty torpedoes. The exposed part of the hull was to be protected by 2in plate over 0.375in shell plating. To navigate the ship a flying deck was provided, connecting two conning towers, and in the early sketches two parts of this light deck were intended to float free and serve as liferafts.

So far so good, but in December 1875 Barnaby wrote to Dunn to inform him that 'it is desired [by the Controller, presumably] to transform the smaller torpedo ship of high speed into a ram of the same general form and distribution of armour'. At some stage the responsibility for executing the design was given to the young Philip Watts.

Only three weeks later Barnaby was writing to the Controller a lengthy version of an unsatisfactory discussion with the aged Admiral Sartorius, the leading advocate of rams. Sartorius wanted an unarmoured ram capable of 15–16kts (fast for the 1870s), with a light sailing rig. If similar to the corvette *Rover*, his ram would cost only £170,000. Barnaby pointed out that HMS *Rover* was probably what the admiral wanted, and showed him a model of the fast armoured torpedo vessel, armed with the unheard-of total of forty torpedoes (although eighteen would be the final total). At this stage Sartorius apparently bowed out, and the proposed ship became primarily a semi-submersible torpedo boat with ramming as a secondary function. The design had the following broad characteristics at this stage of development:

Displacement:	2340 tons
Dimensions:	250ft x 37ft x 24ft
Machinery:	5,000 indicated horsepower
Speed:	17kts

Later it was decided to provide protection for the exposed part of the hull, with 2–3in of compound armour, but after a lengthy series of trials the protection selected was 1in-thick iron

HMS Polyphemus *represented the triumph of fashion over hard-headed thinking about naval warfare, and was rapidly made obsolescent by technological advances.*

plates as an inner layer, and an outer layer of smaller hardened steel plates. She also had a conning tower protected by 8in armour, and the hatch coamings were given 4in armour.

The machinery consisted of two sets of reciprocating engines in separate engine rooms, and two boiler rooms. A most interesting feature was a 250-ton cast iron drop-keel, made in sections, and released by a pair of spindles. Its purpose was to lighten the ship if she sustained hull-damage. She was armed with five torpedo tubes for launching up to eighteen 14in Mk II torpedoes—they had a range of 600 yards and a speed of 18 kts, the same as the ship! The bow tube formed the core of the massive ram, protected by a hinged cast steel cap. A 'two-bladed' bow rudder was fitted to improve manoeuvrability, and could be retracted when not needed. According to reports the bow rudder worked well when the ship was steaming ahead, and reduced the tactical diameter and the time taken to turn a half-circle by about 12 per cent.

Appropriately for such an unusual ship, she was christened *Polyphemus*, after the one-eyed monster of Homer's *Odyssey*, and was the only Royal Navy ship of her generation to be painted grey. She had an efficient ventilating system to offset the fact that in heavy weather she would have to be closed down, possibly for days on end. She was the first ship to have electric lighting. The original idea of Sartorius was to have no gun armament, but the rising menace of the torpedo boat led to her getting six Nordenfelt 1in twin machine guns mounted in small 'towers', on a flying deck to avoid being washed out in a seaway. These were later replaced by single 6pdr (57mm) quick-firing guns.

HMS *Polyphemus* was an impressive ship in many ways, but she was a failure for several reasons. The slowness of the 14in Mk II torpedo has already been mentioned, and it is highly unlikely that she would have been allowed to approach an enemy unmolested. When conceived, the rate of fire of naval guns was very slow, and she might in theory have been able to get close enough to cripple a hostile ironclad, but the development of the quick-firing (QF) gun put an end to such hopes. Although much bigger than contemporary torpedo boats, her low freeboard, combined with her own and an enemy's coal smoke, would have made the task of identifying her target very difficult. Advances in machinery also eroded her original speed advantage.

In short, there were very few tasks in the peacetime navy for which she was suited. Her moment of glory came in the 1885 Annual Manoeuvres, in which Admiral Sir Geoffrey Phipps Hornby and his 'Particular Service Squadron' would attempt to attack a hostile fleet in a defended harbour. To simulate a protected anchorage such as Kronstadt, booms were erected at either end of Berehaven Island in West Cork, Ireland. The main mile-long outer boom consisted of heavy spars held together by a 5in hawser, and protected by four rows of observation mines. There was also an inner boom of lighter spars, and the two were connected, with loose ropes arranged to foul propellers; finally, the booms were covered by field artillery and machine guns.

On 30 June 1885 HMS *Polyphemus* worked up to full speed over two miles, evaded the attacks of six torpedo boats and dodged ten torpedoes, and finally charged the boom. Observers said that she snapped both booms 'like pack thread' and scattered wreckage over a wide area. Aboard the ship no shock was felt, and she suffered no damage at all. But the conclusion drawn from this spectacular event was that ships would not be able to charge recklessly into a fortified anchorage, and it is significant that the concept disappeared from Admiralty war plans very soon afterwards. The power and precision of artillery was improving steadily, and only a miracle would prevent the attackers from being disabled and sunk.

And so *Polyphemus*'s moment of glory was soon forgotten, and although she continued to serve in the relatively placid waters of the Mediterranean, she was finally sold for scrapping in 1903. Two years later, a new revolution would be begun, spearheaded by HMS Dreadnought, designed under the aegis of *Polyphemus*'s own young designer, now DNC.

HMS POLYPHEMUS

Laid down 21 September 1878, launched 15 June 1881, completed September 1882, Chatham Dockyard

Displacement:	2640 tons
Dimensions:	240ft (pp) x 40ft x 20ft 3in
Machinery:	2-shaft compound steam, 7000ihp
Speed:	18kts
Armament:	twin 1in Nordenfelt MGs (later replaced by six 6pdr QF)
	5 14in TT (18 Mk II carried)
Complement:	146

The USN was impressed by the *Polyphemus*, but only for her alleged superiority as a ram. Ramming's great evangelist was Rear Admiral Daniel Ammen, a diehard apostle of a purely coast defence role for the navy. Ammen wanted a fleet of rams, and finally wore down his superiors, who persuaded Congress to authorise the construction of one prototype in March 1889, embodying his ideas. Naval technology was moving at an accelerating pace, however, and objections to ramming valid in 1885 after Berehaven were even more so by 1896, when the USS *Katahdin* was commissioned into service. Her lower hull was dish-shaped and curved upwards at each end, and could be partially flooded to bring the ship to fighting trim. On the sides, below the waterline, the hull had a sharp 'knuckle' which was packed with wood and armoured with 6in plates. The protective plating, varying from 2–6in in thickness, was continued over the deck to form a huge turtleback. The internal structure was framed with longitudinal girders which converged at the bow to support a massive cast steel ram. The deck-fittings and structures were kept to a minimum: an armoured conning tower and funnel, a light mast for signalling, four 6pdr guns in shields, and skid-beams for the boats.

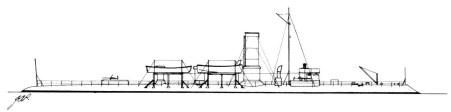

The USS Katahdin entered service at least ten years too late. Technology was already making her obsolete when her authorisation was still being discussed. (Line drawing 1/720 scale.)

Although the *Katahdin*'s engines exceeded their designed horsepower, she was unable to reach her contract speed of 17kts. This caused a legal problem, because the Navy was prevented by law from accepting her. The Navy Department, however, got around the problem by claiming that the unique hullform probably made a speed of 17kts impossible to achieve, and Congress enacted a special bill to allow the *Katahdin* to be accepted.

Life between decks was uniquely uncomfortable, even by late-19th century standards, thanks to narrow passageways with low headroom. Ventilation was bad, and the compartments constantly dripped with condensation. Temperatures of 110°F were regularly recorded in the wardroom, and the heat in the stokeholds was almost beyond human endurance. According to her commanding officer, in any kind of seaway she was 'half-seas under'.

After a year of Navy trials and shakedown, the green-painted *Katahdin* was decommissioned, only to be hurriedly brought back into service at the outbreak of the war against Spain. Her mission was to protect East Coast ports from the nightmare of a Spanish onslaught. In late June 1898 she was at last ordered to Cuba, but while en route she was recalled because of the destruction of the Spanish at the Battle of Santiago. Her big moment had come and gone, and the ram was quickly reduced to reserve once more, this time permanently. Stricken on 9 July 1909, she was designated Ballistic Experimental Target A, and was sunk as a gunnery target off Rappahannock Spit, Virginia two months later.

USS KATAHDIN

Laid down July 1891, launched 4 February 1893, commissioned 20 February 1896, built by Bath Iron Works, Bath, Maine

Displacement:	2155 tons (normal)
Dimensions:	250ft 9in (wl) x 43ft 5in (maximum) x 15ft 1in (mean)
Machinery:	twin screw horizontal triple-expansion steam, 5,000ihp
Speed:	16kts
Armour:	6in on sides
Armament:	4-6pdrs (QF)
Complement:	7 officers, 90 enlisted men

CONCLUSION

The designers and builders of these two warships cannot be blamed for their shortcomings. Rather the opposite, for they achieved remarkable success in meeting very complex requirements, and both ships were well executed. Their failure can only be blamed on the naval authorities who sanctioned them. Had either navy done any analysis worthy of the name, the dubious logic of ramming would have been properly debated. There was ample evidence from deliberate and accidental ramming incidents to show that contemporary steam-driven warships were neither fast enough nor sufficiently manoeuvrable to make such tactics workable.

The slow rate of technical progress of the 'old' US Navy was the result of chronic underfunding, and explains why the *Katahdin* was authorised nearly a decade after the *Polyphemus*. To make matters worse, the inflated reputation of the Civil War monitors created an illusion that heavily armed coast defence ships would be sufficient to preserve the United States from seaborne attack. The rationale of the British ship had been overtaken by the invention of the QF gun, but her American counterpart lacked that excuse. One last objection: the Royal Navy was a very rich organisation, and could afford an aberration or two. The US Navy, in contrast, had to fight for every cent, and a single expensive 'sample' like the *Katahdin* used up funds which could have contributed more cost-effective ships to the nation's naval defence.

It is interesting to note that in 1913 the go-ahead young First Lord of the Admiralty, Winston Churchill, called for a 'new *Polyphemus*', a torpedo-cruiser protected by an arched deck of 3in vanadium steel armour. Armed with below-water torpedo tubes, she (presumably in company with others of her type and destroyers) was intended to close with the enemy battleline and disrupt its cohesion with salvos of torpedoes. The same old objections seem to apply to this design, although it progressed to the stage of a sketch design, but nothing more was heard of the proposal by the outbreak of war in August 1914.

Armoured Cruiser *Rurik*

Imperial Russian Navy 1890–1905

In about 1890 rumours began to spread of a very heavily armed, well-protected and fast cruiser being built for the Imperial Russian Navy. The mystery ship was apparently intended as a commerce-raider, and as the rumours spread the British Admiralty became seriously alarmed. Owning the world's largest mercantile fleet made the British Empire vulnerable to attacks on seaborne trade, and so any increase in the level of threat made the Royal Navy very nervous.

Although privateering had been abolished under the terms of the Treaty of Paris in 1856, the Americans, French and Russians made little secret of their plans to attack British shipping in time of war. During the war against France from 1793 to 1815 the British mercantile marine

The Rurik after the removal of her barque rigging in 1901.

had suffered a 100 per cent turnover of its strength at the start of the war, although access to North American timber and shipyards had ensured that the lost ships were replaced. The advent of steam seemed to complicate the problem, however, and it was widely believed that the trade-protection tactic of convoying vulnerable shipping would no longer suffice. Endurance under steam was low and even when steam machinery became more efficient in the last two decades of the nineteenth century it was feared that fast enemy cruisers would act as 'armadillos', lapping up ants as they emerged from their nest, in the words of Sir John Fisher.

The Russians and the French in particular recognised that their only way of winning a naval war against Britain would be the classic *guerre de course* against commerce on the high seas. Both navies had built a series of nominally unbeatable cruisers, and as it had to be assumed that the Royal Navy would be opposed by both the French and Russian navies acting in concert, the rumours of a new Russian 'paralyser' on the trade-routes fuelled anxiety to the point where it approached paranoia. The British press dutifully took up the call, and 'naval experts' predicted disaster on the high seas.

What finally appeared in 1895 was a blow to the myth-makers. The *Rurik* turned out to be a barque-rigged large armoured frigate with her armament on the broadside. Far from being ultra-fast, she was only good for 18.7kts in light condition. The armament was heavy, four single short 8in 35-cal guns sited in sponsons port and starboard at the break of the forecastle and at the poop, protected by small shields, 16 single 6in 45 quick-firers on the main deck, unprotected, and six 4.7in on the upper deck, protected by shields. There were also four 15in torpedo tubes above water, port and starboard.

The main protection comprised a 10–5in belt of Creusot steel, 6ft 9in deep, terminated by 10in bulkheads taken up to the upper deck to protect the 6in guns from raking fire. There was also a 2.5in deck, thickened to 3.5in forward and aft. The belt armour was also augmented by coal, disposed above the armoured deck and subdivided by transverse bulkheads. Her hull was sheathed and coppered to avoid fouling from marine growths on long voyages; the barque-rig was chosen to extend endurance when cruising in distant. waters.

The powerplant comprised four vertical triple-expansion reciprocating engines driving twin shafts. Steam was provided by eight cylindrical boilers, and the maximum coal-stowage was 2000 tons. The barque-rig was later reduced to a barquentine-rig, with square rig on the foremast only, and two lighter 'military' masts as main and mizzen, but nothing could be done about the antiquated broadside arrangement of the armament.

The design originated from a staff requirement for a long-range commerce raider sent by Admiral I A Shestakov, Director of the Naval Ministry, to the state-owned Baltic Works in St Petersburg, although the outline specification has not been found in the Russian archives. It seems that Shestakov wanted an improved version of the belted cruiser *Pamiat Azova*, then being built at the Baltic Works. He emphasised long range, presumably to permit the ship to sail from the Baltic to Vladivostok without coaling *en route*, as well as good seakeeping.

Shestakov deliberately avoided using the *Morskoi Technicheskii Komitet* (MTK) or Naval Technical Committee as he did not trust it to produce ships of the optimum size and cost. In fact, throughout his tenure he bypassed the MTK and its members, not least because they had worked for his predecessor. One of the shipyard's constructors, N E Rodionov, prepared a design displacing 9000 tons and protected by an 8in belt, resembling the *Pamiat Azova*. It was submitted to the Naval Ministry in July 1888 and then to the MTK three months later for review. The MTK was resentful of Shestakov and of the head of the Navy, the General Admiral (the Tsar's younger brother, the Grand Duke Aleksei Aleksandrovich). The result was very severe criticism of Rodionov's design.

The design had been given a high sustained speed and large coal-stowage by adopting a long and narrow hull (420ft x 61 ft on the waterline). The MTK attacked the design on the grounds that the length was excessive, which made for lack of manoeuvrability. It also objected to the length:beam ratio of 6.8:1 as too fine for a warship. The Baltic Works team pointed out that the design lacked heavy gun turrets forward and aft, so a fine hullform was not inappropriate. The MTK also questioned the claimed endurance of 20,000 nautical miles, saying that 15,000 nautical miles was more likely, and suggesting that Rodionov had omitted certain weight factors from his calculations. Accordingly a 10,000-ton design was preferred, but in November 1888 Shestakov died. His successor, Admiral N M Chikhachev, enjoyed more harmonious relations with the MTK; the Baltic Works design was rejected by the General Admiral, and work began on a new MTK design.

A new set of requirements was worked out for a cruiser with speed reduced to 18kts, 10,000 nautical miles endurance and the same armament as before. A barque rig was mandatory, and a 10in armour belt (the MTK did not have much faith in a protective deck). Constructor Kuteinikov produced a 9000-ton design and a 10,000-ton one, but the latter was accepted as superior. The design was ready in July 1889, but a sudden problem arose over the powerplant. Originally it was hoped to adapt the design of the Napier machinery in the battleship *Sinop*, but when these failed to reach their designed speed the idea was dropped. Instead Kuteinikov recommended the idea of coupling two engines to each shaft, but a proposal to adopt the advanced French Belleville boiler was rejected in favour of the older and less efficient cylindrical type.

The Imperial Russian Navy was notorious for its habit of adding 'improvements' after the design was completed, resulting in an increase in displacement. The new cruiser, to be named *Rurik* after the semi-mythical ninth century ruler of Kiev, was no exception. Among the numerous alterations was the substitution of 6in 45-cal guns in place of the 6in 35-cal weapons, and 4in transverse bulkheads at the end of the battery deck. Other weight-increases were caused by the provision of two 14-ton 2nd Class torpedo boats, rather than the normal steam launches, and 95 tons of steel to strengthen the hull. Despite being 633 tons overweight she managed to make a maximum of 18.8kts on trials, and developed 13,362 ihp.

In service the *Rurik* proved a good steamer and could fight her guns in a seaway. She was sent to the Far East, to the main Pacific Squadron base at Vladivostok. Only six years after completion, Admiral Dubasov, commanding the squadron, recommended the complete removal of her heavy masts and yards, and the replacement of the inefficient cylindrical boilers with Bellevilles, but neither change was approved. The dockyard facilities at Vladivostok were not considered to be capable of a reboilering, although modifications to the rig were presumably vetoed for other reasons; it was hardly a major burden on the dockyard's resources. Instead her sail area was reduced from 28,000 sq ft of canvas to 6800 sq ft, the bowsprit and topgallant masts were removed and the rig was cut down to a barquentine style, with crossed yardarms on the foremast.

The ultimate test of a warship's qualities is battle, and for the *Rurik* the test was to prove fatal. When war broke out with Japan early in 1904 she was at Vladivostok with the more modern armoured cruisers *Rossia* and *Gromoboi* and the large modern protected cruiser *Bogatyr*. The main mission of the squadron was to attack Japanese shipping and to intercept the traffic between Japan and the main theatre of war, Manchuria.

By August 1904 the squadron had made six sorties, but the only result was the sinking of the transport *Hitachi Maru* in June. That single ship was, however, carrying 18 11in howitzers intended for the siege of the main Russian base at Port Arthur. These siege howitzers were replaced by early August, and they began to shell the Russian battleship squadron, forcing the high command to order a breakout to Vladivostok. The sortie on 10 August was frustrated by the Japanese blockading squadron, and most of the force returned ignominiously to Port Arthur, with the exception of the light cruiser *Novik*, which was forced to seek internment in a neutral harbour. The cruisers at Vladivostok were intended to support the breakout but the orders were delayed, and arrived when the boiler fires were drawn. As a result the ships could not sail until 13 August, by which time the main fleet was back in Port Arthur. The *Bogatyr* had been damaged by grounding, so she was left behind.

Admiral Iessen's ships were north of Tsushima at first light on 14 August when four Japanese warships were sighted. They were a squadron of armoured cruisers under the command of Vice Admiral Kamimura, who positioned his ships to cut off the Russians' line of retreat. At 05:00 Admiral Iessen, flying his flag in the *Rossia*, ordered a change of course to the northeast, and his ships opened fire at 05:23 at a range of 8500m. Kamimura's ships followed suit, but on a slightly converging course. Within half an hour the *Rurik* was hit and began to drop back.

At abut 06:00 Iessen ordered a 180 degree (i.e. reciprocal) turn, hoping perhaps to open his ships' arcs of fire or to allow the *Rurik* to rejoin the squadron. But the Japanese ships could now 'cross the T' and rake the Russian line. A few minutes later Kamimura's flagship, the *Idzumo* also made a reciprocal turn, but as it was to port the range opened. He was forced to slow down when the *Azuma* developed machinery trouble, and fire was not resumed until 06:24. *Rurik* suffered three more hits on her stern; her tiller flat flooded, putting her rudder out of action, although she was still able to steer on her engines. Speed dropped steadily, making her more vulnerable to Japanese fire, and at about 06:40 her rudder jammed, making her unmanageable, circling slowly to starboard.

At this juncture Iessen ordered a 180 degree turn to port, placing the *Rossia* and *Gromoboi* between Kamimura's ships and the hapless *Rurik*. Suddenly the apparently crippled cruiser increased speed and swung to starboard, passing between the flagship and the Japanese squadron. After two more attempts to rescue her, Iessen took the other cruisers northwards, leaving the *Rurik* to her fate. She was attacked by the protected cruisers *Naniwa* and *Takachiho* under the command of Admiral Uriu and fought from 08:42 until 10:05, by which time she was clearly doomed. Her senior surviving officer, Lieutenant Ivanov, ordered the wounded to be laid on deck and ordered the torpedo magazine to be detonated. No fuses could be found, so the Kingston valves in the engineroom were opened, allowing her to sink by the stern. The crew tied the wounded to life rafts or pieces of timber and then abandoned ship. In a calm sea the Japanese rescued 625 survivors, including about 230 wounded; some 170 of *Rurik*'s crew went down with the ship.

RURIK

Laid down 19 May 1890, launched 22 October 1892, commissioned May 1895, built by Baltic Works

Displacement:	10,993 tons (designed), 11,960 (as completed)
Dimensions:	412ft (pp), 435ft (oa) x 67ft x 26ft
Machinery:	2-shaft VTE (2 on each shaft), 13,250ihp (designed);8 cylindrical boilers
Speed:	18kts (designed)
Armament:	4-8in 35 cal (4 x 1), 16-6in 45 cal (16 x 1), 6-4.7in 45 cal QF (6 x 1), 6-47mm 43 cal (6 x 1), 10-37mm 23 cal (10 x 1), 6–15in torpedo tubes (AW)

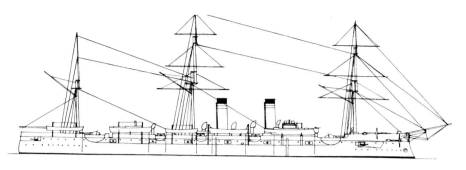

The Rurik *appeared antiquated even on her commissioning in 1895. She fought gallantly at Ulsan, against superior ships, but never really fulfilled her role of commerce raider. (Line drawing 1/1250 scale.)*

Armour:	8–10in belt, 2–3in decks, 6in CT
Coal/endurance:	1933 tons (maximum)/6700nm @ 10kts
Complement:	719 officers and ratings

Conclusion

The *Rurik* is an unusual example of a ship dismissed by most naval historians simply because she looked obsolete. The square rig, lack of protection for her guns, and their broadside disposition marked her as old-fashioned, a problem exacerbated by her lengthy building time.

In terms of how she lived up to her designers' expectations, however, she could hardly be called a failure. She made her designed speed with a margin to spare, and as a long-range commerce-raider she was never designed to fight her fellow armoured cruisers. The essence of commerce-raiding was to provide sufficient speed to catch merchantmen, enough range and seakeeping to operate in mid-ocean, and sufficient armament and protection to defeat or frighten off small cruisers. In that context, sheer fighting power against large armoured cruisers was irrelevant. There is, of course, a drawback to this policy: given the unpredictability of war, such specialised ships may end up fighting a conventional battle. This is what happened to the *Rurik* at the Battle of Ulsan on 14 August 1904.

For a ship damned as obsolete, she put up a very good fight against Kamimura's modern ships. She fought for nearly five hours, during which time she suffered a number of 8in shell hits, although there is some suspicion that the Japanese shell-fuses and fillings were not reliable. Despite being overwhelmed and out of control she was not sunk from enemy action, but was scuttled after every effort was made to save life.

Analysis of the battle shows that the Japanese should have sunk all three Russian cruisers, but Kamimura opened the range at critical moments and seemed unwilling to close to decisive range. Iessen's ships fought well, but their gunnery was inferior, scoring only 2.5 per cent hits as against 6 per cent Japanese. His loyalty in trying so hard to save the *Rurik* was doubtless good for Russian morale, but as Peter Brook has pointed out, he should have abandoned her earlier.

The *Rurik's* sacrifice and the Battle of Ulsan as a whole had no decisive effect on the outcome of the war. The *Gromoboi* was soon out of action as a result of grounding, and the *Rossia* led no more sorties. Kamimura was lauded for his victory but it was flawed by a surprising degree of indecisiveness and his failure to catch the fleeing *Gromoboi* and *Rossia*. He was also very lucky not to lose the *Iwate* to a cordite fire in a casemate (which left 41 dead and 34 wounded).

DYNAMITE CRUISER USS *VESUVIUS*

US NAVY 1890–1922

In 1886 a new and unorthodox type of weapon made its first appearance, the smooth-bore 'dynamite gun'. Dynamite, produced by Alfred Nobel for industrial blasting, was well-known, and it had far more destructive power than the traditional 'black' gunpowder in use by navies and armies world-wide. But dynamite was too sensitive to withstand the shock of being fired in a conventional shell, making it potentially capable of damaging or even bursting the gun and, incidentally, killing the gun crew. The only other explosive available was guncotton, but this also generated too sharp a shock, and its use was restricted to mines and torpedoes, principally because it was not affected by seawater.

An enterprising young US Army officer, Lieutenant Zalinski, turned his mind to the problem in the early 1880s. His solution was a guncotton-filled projectile expelled from an unrifled barrel by compressed air, creating a much slower build-up of pressure. He convinced some rich backers, who formed the Pneumatic Dynamite Gun Company of New Jersey to market his invention. Despite its name, the gun was intended to fire guncotton shells.

Influential senior officers in the US Navy liked the idea too, and on 3 August 1886 Congress authorised the expenditure of $350,000 for the construction of an experimental 'dynamite cruiser'. The Pneumatic Dynamite Gun Co. was awarded the contract, and it in turn awarded a building contract to William Cramp & Sons of Philadelphia.

Although the new ship, aptly named *Vesuvius*, was rated as a cruiser, she had more in common with an elegant steam yacht. The 252ft hull was totally unprotected, but she was fast by the standards of the day, being capable of a speed of 21kts. In one respect, however, she was a disappointment. As a result of a quirk of the after hull form, she had the widest turning-circle of any contemporary US Navy warship. This was a serious shortcoming in a ship whose survival depended on speed of attack and manoeuvrability in the face of hostile gunfire.

By far the most imposing feature of this little ship was her battery of three 'dynamite' guns, 15in in calibre and 54ft long, in a recessed position on the forecastle. The breech of each gun was below the waterline, and was reloaded from one of three revolving magazines, each containing nine 'dynamite' shells. The guns themselves were controlled by complex machinery and valves, to allow firing by admitting air from giant cast-iron air storage flasks. The pressure in these flasks was 1000lbs per in^2.

A major drawback was the fact that the *Vesuvius* had no provision for training or elevating the guns. Aiming had to be achieved by turning the ship herself, and her erratic steering made

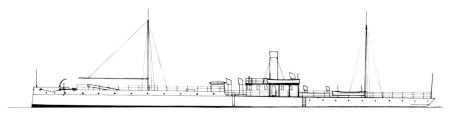

The USS Vesuvius *was more of a gun with a ship attached, than a naval vessel. (Line drawing 1/720 scale.)*

this even more of a problem when trying to hit the target. The range was adjusted by varying the duration of the air blasts, or by varying the weight of the projectile. Three sizes of shell were available. The largest was nearly 7ft long and contained 500lbs of guncotton. Two smaller types were used when increased range was needed; they were sub-calibre rounds fitted with wooden sabots to fit the 15in barrels. As the barrels were smooth-bore to reduce friction the shells were fitted with stabilising fins. Varying the projectiles and the compressed air blast gave ranges varying from 500 yards out to 2000 yards.

Firing the guns was virtually silent, but the effect of the projectiles was very impressive. During trials on the Delaware River in 1890 the *Vesuvius* fired a number of dynamite shells which created columns of mud and water and rattled windows up to five miles away, according to contemporary accounts. Foreign navies and distinguished civilian commentators were suitably impressed, giving the new ship credit for a lot more than she could ever achieve. The idea of a silent approach, followed by a silent discharge of weapons seemed to herald a new type of tactic which would be impossible to counter. Today we can only imagine the effect of a discharge of a battleship's heavy guns; at night the flash dazzled anyone trying to direct fire, gunners suffered from being too close to the blast, and the noise was literally deafening. In daylight the flash of guns firing was not significant, but the effects of blast and noise were still a problem to a generation of sailors who knew nothing of ear-defenders and would have regarded them as effete. To eliminate all these factors simultaneously seemed to offer so many advantages that assessments of the dynamite cruiser's other qualities took much for granted. By contemporary standards she was fast, and slow-firing, medium- and heavy-calibre guns might find her a difficult target.

When she was commissioned the little cruiser attracted much public interest for her perceived qualities. Her shallow draught permitted her to visit many harbours not visited by larger ships, and a wider section of the American public became aware of her existence as a result. The Navy, however, became aware of her deficiencies very quickly. The greatest problem was to achieve consistent ranges, because the firing valves had to be accurate to within thousandths of a second, something beyond contemporary ordnance technology. The shells also had unreliable fuses, and firing trials yielded an unacceptable proportion of 'duds'. By the 1890s many naval officers were aware of the damage that would be inflicted on the unarmoured hull of the *Vesuvius* by the latest quick-firing guns. There was a great risk of her being crippled or even destroyed before she could bring her dynamite guns to bear, if her magazine detonated, particularly if her manoeuvrability was affected by poor steering.

The Navy planned a second dynamite cruiser as soon as the trials of the *Vesuvius* were completed. Using funds allocated in Fiscal Year 1890, the unnamed ship would have been armed with only two 15in dynamite guns and would also have been capable of 21kts. No design work was done, however, and in July 1894 Congress transferred the funding to three torpedo boats.

The Pneumatic Dynamite Gun Co. was sliding into bankruptcy by the mid-1890s, and the Navy's refusal to spend money on a second vessel of improved *Vesuvius* type was the last straw.

The concept was now discredited, and stocks of ammunition were expended without being replaced. As a final humiliation, there was talk of converting her to a torpedo boat.

The sudden outbreak of war between the United States and Spain over the Spanish colonies in the Philippines and Cuba gave the *Vesuvius* a brief reprieve from oblivion. In June 1898 she was sent to join the Blockading Squadron off Santiago. As predicted by her critics, she stayed out of harm's way during the hours of daylight, but at night crept inshore and lobbed shells at the Spanish fortifications. Sailors in the covering ships were intrigued by what contemporary sources called a noise 'like a giant's cough' in the darkness as she fired her pneumatic guns. But it was only random firing, with a few shells falling harmlessly among the Spanish warships in Santiago Harbour, and others blowing holes in empty hillsides. Undoubtedly they frightened the defenders, but as Secretary of the Navy John D Long commented after the war, 'the final effect was materially unimportant though morally great'.

Surprisingly, after such a damning comment, the *Vesuvius* survived for another quarter of a century. From 1905 to 1921 she served in the humble role of an experimental torpedo-firing tender. She was downgraded to 'unclassified' status on 17 July 1920, put up for sale on 13 October 1921 and finally sold for scrapping on 19 April 1922. Hardly a short life, especially for a ship whose faults were all too evident a quarter of a century earlier.

USS *Vesuvius*

Laid down September 1887, launched 28 April 1888, commissioned 7 June 1890

Displacement:	929 tons (normal)
Dimensions:	252ft 4in (oa) x 26ft 6in x 10ft 9in
Machinery:	2-shaft vertical triple-expansion steam; 3795 ihp
Speed:	21.42kts
Armament:	3-15in pneumatic guns; 4-6pdr QF
Complement:	6 officers, 64 enlisted men

Conclusion

The *Vesuvius* was a failure in that she was not the 'breakthrough' that her advocates claimed, but she had many novel features, and it is to the credit of the US Navy that the design was given a fair trial. To a modern reader, the most interesting point about her must be the fact that she was an early example of a 'complete weapon system', a ship designed around a specific weapon, with all ship-features subordinated to the planned mission. There is one unsolved minor mystery. Why was she persistently described as a 'dynamite' cruiser when all surviving records show that her projectiles were filled with guncotton? This was a well-known explosive, so the reason cannot be secrecy. Presumably the image of a 'guncotton cruiser' lacked the glamour of a link with the latest and most destructive explosive on the market.

In other important ways she was a failure, however, but not all of this was the fault of the designers. The concept of the dynamite cruiser was overtaken by the rapid advance of tactics and weaponry in the late 1880s and early 1890s; daylight attacks would have been suicidal in the face of the new generation of quick-firing, light- and medium-calibre guns proliferating in all major navies. By the time of the Spanish-American War this vulnerability was obvious, hence her restriction to night bombardments. She was also too small for open-sea operations, and she was essentially suited to coastal waters, a late variation on the monitor theme. The biggest failure of all, however, was the dynamite gun, which failed to match the accuracy or reliability of conventional heavy ship-guns. The technology did not yet exist to provide such accurate aiming, but we must also question the faulty fuses of the projectiles. Contemporary accounts do not explain if this was a problem of quality control, but that is the most likely culprit. Not for the first time, a major weapon system was rendered less effective by minor sub-system failures.

The designers were also trapped by some inconvenient laws of physics and hydrodynamics. To give the dynamite cruiser any degree of protection against enemy gunfire she would have had to be much bigger, and if the designed speed was to be achieved her machinery would have to be more powerful. These criteria could not be reconciled in a small hull as contemporary steam machinery was heavy and had a relatively poor power:weight ratio when used in small hulls. The designers of HMS *Polyphemus* and the USS *Katahdin* (see page 28) were caught in a similar trap; they could have speed or protection, but not both.

To sum up, the *Vesuvius* was a worthy experiment, but in today's terminology, her weapon system was not properly 'derisked'. The pneumatic gun should have been subjected to a lengthy series of trials on land before the ship's design was finalised. If the trials had been properly supervised by Navy ordnance experts most if not all of the faults would have been detected. This was not the only example of too much haste to get the ship built, and for that the blame must rest partly on the shoulders of the influential officers who persuaded their fellow-officers and Congress to fund the project. Money was tight, and shore trials would have added to the total cost, but a more likely reason is the very human conviction that no time should be wasted in giving one's Navy the benefit of a major advantage over its competitors—in this instance the British, French and Russians. Nor would such an achievement do harm to the professional careers of the project's supporters if it turned out to be a great success. Vanity and the desire for self-advancement are very human failings, and navies are made up people as well as ships.

POWERFUL CLASS PROTECTED CRUISERS

ROYAL NAVY 1894–1932

The Royal Navy's fundamental and unchanging mission in the 19th century was the preservation of the Empire, a role which included defence of the British Isles and the trade links with the Empire, on which British commercial supremacy depended. For that reason the Royal Navy built a wide range of cruisers in the second half of the century, varying in size from large protected and armoured cruisers (not the same thing) through corvettes, 2nd and 3rd class cruisers, some really large sloops, and fast but unprotected scouts.

Originally there was no separate cruiser category; 'cruizing' in the age of sail merely indicated a ship operating independently while protecting trade or harrying enemy shipping. But by 1860 it was emerging as a category outranked by the line-of-battle ship but superior in gunpower to all other ships. One distinguished scholar has described the cruiser as the largest type of warship which could be built in numbers, although that was hardly true of some designs.

Compared with the development of the ironclad, development of the corvette and sloop in the 1860s and 1870s was slow. In general they were a continuation of types dating from the sailing navy, and improvements were concentrated on machinery and armament. Serving on foreign stations required range, and with a shortage of coaling stations auxiliary sails were essential, especially in the Pacific. The chain of dockyards had not yet been established either, so wooden hulls, normally sheathed with timber and coated with thin copper to resist the depredations of the teredo ship-worm and to inhibit marine growths such as barnacles, were preferable to iron ones.

The limiting factor in nearly all these cruising ships was slow speed, although the big iron frigates *Inconstant*, *Raleigh* and *Shah* and the iron corvettes *Active* and *Volage* were faster. The true cruiser did not begin to emerge until 1874, when the Director of Naval Construction, Nathaniel Barnaby, was given the task of producing solutions. The process started with the 'despatch vessels' *Iris* and *Mercury*, built at Pembroke in 1875–79. They were the first British warships built of steel, as opposed to wrought iron, and *Iris* caused a sensation when she achieved a speed of 18.6kts on trials.

Despite their innocuous designation as despatch vessels (they were originally described as fast corvettes), the *Iris* and *Mercury* were almost certainly intended to hunt French commerce-raiders. They carried a heavy armament of ten 64pdr rifled muzzle-loaders and stowed a lot of coal. Although unarmoured, they were well protected, with two boiler rooms and two sets of engines below the waterline, and deep coal bunkers abreast of them preserved buoyancy and stability in a damaged state.

Their successors, the *Comus* class corvettes, seemed a step backwards, but they adopted a steel protective deck 1.5in thick. It was much more than a simple armoured deck as it formed part of a comprehensive system, with deep bunkers affording additional protection against shellfire. The deck was about 3ft below the waterline, affording protection against anything except high-angle fire from coastal batteries.

Sir William White took Barnaby's concept and developed it, creating a series of 1st, 2nd and 3rd class protected cruisers for guarding the trade routes. There was some ill-informed criticism about the cruisers White designed for the Royal Navy after he left Armstrong's Tyneside yard, but many of the superior features ascribed to his 'Elswick' export cruisers do not stand up to close scrutiny.

Then came the *Rurik* scare (see page 33), and a widespread conviction that no existing British 1st class cruiser could catch her or face her in battle. In response to agitation for a 'paralyser' to match the Russian cruiser, funds were voted for two high-speed, long-range protected cruisers. They were also to have a secondary function as fast transports to move troops to trouble-spots. White gave the Admiralty what it asked for, the largest and longest cruisers in the world, with a powerful set of engines for high speed and massive coal capacity for range.

The dimensions of the new cruisers were largely determined by the size of available dry docks, for there were few in the British Isles or at dockyards around the Empire long enough. There were even fewer docks with the width needed to match the beam necessary to maintain stability. Some Admiralty Board members were uneasy about the decision to build such huge ships, and suggested that length should be cut to 450ft. White responded somewhat tartly that the machinery spaces would take up the same length, that a shorter hull would make even 20kts difficult to maintain, and that a length of 450ft would not solve the docking problem.

HMS Terrible, *and her sister* Powerful, *were expensive ships designed to fight a specific enemy. The fact that they never met this foe raises questions about designing ships so specifically.*

HMS Powerful *after her funnels had been raised in 1897 to improve draught for the boilers. They were handsome ships, at least.*

The original armament envisaged was a uniform battery of twenty 6in guns. The Director of Naval Ordnance (DNO) opposed this scale of armament with great vigour. He pointed out that the most important positions in the ship were the forecastle and the poop, where 6in guns would be inadequate as they lacked hitting-power. In his opinion the rate of fire of a 6in gun counted for less than the potential damage inflicted by a heavier shell, and advocated an 8in calibre main armament. He was sure that Armstrongs could produce a new twin 8in 40-cal gun, and proposed mounting fourteen or sixteen 6in guns in light shields.

White fought his corner well, pointing out that a 6in armament was adequate for fighting other cruisers, as live firings against the target-ship *Resistance* had shown, whereas if the new cruisers were intended to mop up crippled battleships after a fleet action, the 8in lacked hitting-power. He proposed the 9.2in gun as the smallest calibre for the job, and eventually won the day, although eight 6in guns had to be sacrificed to offset the greater weight. The 'bogey' cited was the USS *Brooklyn*, which it was feared would be able to crush her pursuer with 8in shell-fire, at a range at which the British cruiser could make no reply. The question of how two ships running at high speed would be able to score repeated hits with the primitive fire control of the day was left unanswered. In any case, ships driven hard on reciprocating steam machinery were subject to severe vibration, a further hindrance to good target-practice.

The ships started trials in 1896, but several problems developed with machinery. In part these were caused by lack of familiarity with the Belleville boiler (they were the first large Royal Navy ships to adopt water-tube boilers). The Admiralty had specified cylindrical boilers, but the Engineer-in-Chief, Sir John Durston, encouraged by White, substituted Bellevilles, although some favoured a compromise, with a mixture of cylindrical and Belleville boilers. This argument was met by the criticism that stokers would have to train on two different boilers. The triple-expansion engines also gave trouble, and commissioning was delayed for over a year. On trials the *Powerful* reached 21.8kts with 25,886ihp, while her sister reached 22kts with 25,572ihp. They continued to suffer from mechanical problems throughout their careers, but could maintain high speed for long periods. As the engineers learned more about the quirks of the Belleville boiler the complaints about the design decreased.

As the first four-funnelled cruisers in the Royal Navy, the *Powerful*s looked impressive, and the raising of their short funnels in 1897 to improve draught for the boilers made them look even better. In 1902–04 both ships had the four single casemates on the main deck replaced by double-storey casemates, adding four more 6in guns at upper deck level and matching the casemates forward and aft.

The design included an extra deck, as compared to earlier 1st class cruisers, although this 'boat deck' had a large well amidships around the funnel uptakes and engineeroom hatches. The crown of the protective deck was 3ft 6in above the waterline, and its outer edge 6ft 6in below the load waterline. The deck was 6in thick over the machinery, but thinned to 2in forward and 3in aft, with 4in over the magazines and 2.5in over a narrow section of the crown. Although most references suggest that all guns were fully protected in turrets or casemates, the Mk VIII 9.2in guns were given open-backed shields.

The slopes of the protective deck were 4in thick, but the bottom 1ft, where the frames passed through, was only 2.5in thick. It was reckoned that in the event of a lucky shell penetrating this narrow strip of armour it would still have to penetrate the whole depth of the coal bunker. They were the first ships with an ammunition passage below the protective deck. This avoided the need to stow ready-use ammunition in the casemates, a very real danger to the ship's safety.

Sir William White took a great personal interest in the construction of both ships, and visited the major shipyards to satisfy himself that they were capable of building 500ft long hulls. Many would need to extend and strengthen the slipways, and there were doubts about the finances of Palmers at Jarrow and Earle's at Hull. White had immense experience of commercial shipyard practices, and had transformed the efficiency of the Royal Dockyards; he was probably the best person in the country to assess the yards' competence.

The two cruisers were criticised for not having sufficient armament, but as White pointed out, the existing hull could accommodate neither the extra magazines nor the gun crews to operate them. They were already very big hulls, and would have needed to be even larger.

The high freeboard made for excellent seakeeping, and the ships were assigned to the China Station when first commissioned. They caught the public's imagination in 1899, when they landed naval brigades in South Africa and played a part in the Relief of Ladysmith. They then re-embarked the naval brigades and rushed them to China to help suppress the Boxer Rebellion. They were laid up in reserve after their 1902–04 refits, and although still on the Navy List when war broke out in August 1914, played no part in the active fleet. In 1915 *Powerful* was recommissioned as a training ship for stokers at Devonport. In November 1919 she was renamed *Impregnable II*, a role in which she continued until 1929. In 1915 *Terrible* was stripped of most of her armament and converted to a troop transport taking drafts to Egypt for the Dardanelles Campaign. From 1916 she served as an accommodation ship at Portsmouth. Reduced to harbour service in January 1918, she became the training ship *Fisgard III* in August 1920 and served until 1932.

POWERFUL CLASS

Powerful laid down 1894, launched 24 July 1895, completed 8 June 1897 by Vickers at Barrow in Furness

Terrible laid down 1894, launched 27 May 1895, completed 24 March 1898 by J & G Thompson on Clydebank

THE WORLD'S WORST WARSHIPS

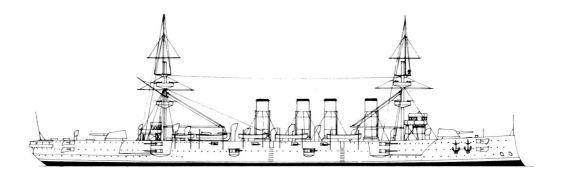

HM Ships Powerful *and* Terrible *were longer than some contemporary battleship designs. (Line drawing 1/1250 scale.)*

Displacement:	14,200 tons (deep load)
Dimensions:	500ft (pp), 538ft (oa) x 71ft x 27ft
Machinery:	2-shaft 4-cylinder triple expansion, 25,000ihp 48 Belleville boilers
Speed:	22kts
Armament:	2-9.2in 40-cal MK VIII (2x1), 12-6in 40 cal QF Mk I/II (16x1), 16-12pdr QF (16x1), 12-3pdr (47mm) QF (16x1), 4 torpedo tubes (1 bow, 1 stern, 2 beam, all submerged)
Armour:	2-6in deck, 6in barbettes, 6in main armament shields
Coal/range:	3000 tons/7000 nm @ 14kts
Complement:	894 officers and ratings

CONCLUSION

The reception given to these giant cruisers was critical from the moment they appeared. The main complaint was their apparently weak armament on such a large displacement, although this was hardly fair, given the overriding requirement for high sustained speed and range. In fact, the DNC's Department issued a formal rebuttal to trenchant criticisms in *The Engineer*. It was pointed out that armament accounted for 27 per cent more of the displacement than in the preceding *Edgar* class, and was protected by 660 tons of armour, as against 340 tons in the *Edgars*. White also pointed out that more guns used up more ammunition, and as it was impossible to enlarge the magazines, the fitting of more guns was pointless.

The Admiralty certainly did not regard the ships as satisfactory. They required a crew 64 per cent bigger than the *Edgar* class, and cost 61 per cent more, while mounting only four more 6in guns. The extent of the discrepancies was even more embarrassing when the true characteristics of the *Rurik* became known. It was obvious to everyone that the ships would not have been built if accurate intelligence had been available earlier. They were also very expensive to run, and it is not surprising that their active careers spanned only seven or eight years.

The *Powerful* and *Terrible* also mark the most extreme manifestation of the folly of building cruisers to match specific opponents. In practice the intended opponents never meet, and it is always wiser to build affordable ships in large numbers. No navy could afford to build large numbers of the *Powerful* type, and even the British Empire at the height of its power found them too expensive.

The saga of HMS *Powerful* and HMS *Terrible* is an object lesson in the dangers of accepting over-enthusiastic intelligence estimates. Such errors were common during the Cold War, when the 'Intelligence Community' made some wild assessments of the Soviet Navy's capabilities.

Serious over-estimates of the alleged superiority of its warship-designs were justified by airy claims that Western intelligence officers had a duty to 'worst-case the scenario'. The usual cause is assessments made by senior officers lacking the requisite degree of technical knowledge, although Sir William White played his part by over-estimating the maxim speed of the *Rurik* – a rare lapse.

It must be remembered that the 1890s were a decade of international tension, and the Royal Navy was particularly obsessed by the fear of a joint France-Russian onslaught against British shipping. In such a climate bad news travels faster and further than balanced assessments of probabilities – the recipients get the intelligence they want to hear.

For all the criticisms, the *Powerful* and *Terrible* represent the high point of protected-cruiser development, even if they had no place in the Royal Navy's inventory. One more group of large protected cruisers, the 11,000-ton *Diadem* class, was built, but they too had short service-lives before being relegated to subsidiary duties. Advances in metallurgy soon made belt armour attractive once more (the steel was lighter but had more resistance), and big armoured cruisers became fashionable by the end of the century.

In spite of all their drawbacks the *Powerful* and *Terrible* were a remarkable technical achievement, unmatched by any contemporary cruiser. The water-tube boiler was the way ahead, even if the Belleville was soon overtaken by improved types. The layout of single 9.2in guns and 6in casemate guns between became standard for the new generation of armoured cruisers. White elephants they may have been, but magnificent white elephants all the same.

BORODINO CLASS BATTLESHIPS

IMPERIAL RUSSIAN NAVY 1899–1922

The *Borodino* class must rank as the unluckiest group of battleships of all time, with four out of five lost. They were also an unsatisfactory design, an improvement over a flawed design—but with several additional vices.

The genesis was the *Tsessarevitch*. Ordered in 1898 from the French La Seyne yard in Toulon, she reflected contemporary French practice, with high freeboard and pronounced 'tumble-home'. (A hull side courving inward from the waterline so that the upper deck's beam is markedly narrower than at the waterline.) The forecastle deck extended right aft as far as the mainmast, giving her a towering, not to say menacing, profile. The secondary armament of six twin 6in guns in turrets was disposed in broadside position, two abreast of the forward super-structure. Two abreast amidships at a slightly lower level, and two abreast of the after super-structure on the same level as the foremost pair.

The armouring scheme was complex. A complete 7–10in waterline belt extended 5ft below the load waterline and 7ft above it, in theory. The lower strake was 10in amidships, tapering to 7in at the lower edge and 8ft at the upper edge. Forward and aft the lower strake thinned to 6in and 6.75in respectively; the upper strake thinned to 5.75in and 4.75in respectively. The turrets and the crown of the conning tower were protected by 2.5in. There were two armoured decks, the main deck being 2.5in over the belt, while the lower deck was 1.5in thick. The latter was curved down near the sides to form a torpedo bulkhead extending from forward of 'A' turret aft to just abaft 'Y' turret. This protection against underwater damage was based on French Navy trials at Toulon in 1890, but was inadequate at it was only 6.34in inboard.

The *Tsessarevitch* was torpedoed in the Japanese surprise attack on Port Arthur on 9 February 1904, but the hit was too far aft to test the torpedo bulkhead. In the Battle of the Yellow Sea on 10 August 1904 she was hit by a 12in shell on the bridge; this shell killed Admiral Vitgeft and another 12in bursting on the sighting slit on the conning tower killed or wounded all the occupants. At this crucial moment the helm jammed, putting her out of control. The stricken ship escaped but was forced to accept internment in the German port of Kiao-chau (Tsingtao); she had survived thirteen 12in hits.

Part of the Baltic Fleet in the First World War, she engaged the powerful German dreadnought *Kronprinz* in Moon Sound on 17 October 1917, but escaped with only damaging hits. In 1917 she was renamed *Grazhdanin* (citizen) to reflect the aspirations of the revolutionary regime, but was laid up and finally scrapped in 1922.

The Tessarevitch *was built by French shipyards for the Tsar's navy and served as a model for the Borodino class.*

The Russians followed the *Tsessarevitch* with a modified design almost immediately, but alterations to the armour-scheme were unhelpful. The Krupp's cemented (KC) armour belt was thinned and made shallower by 1ft. The 3in side armour between the main and upper decks, added to protect the 11pdr battery, was of little value, serving only as a 'shell-burster'. The upper deck was 2.5–1.5in thick over this battery, the main deck was 2in thick and the lower deck 1.5–1in. The torpedo bulkhead was thinned to 1.25in, but unlike the *Tsessarevitch*, was no longer formed by curving the lower deck armour down. Instead, it was joined to the deck by a narrow void space or 'flat'.

As the latest and most powerful battleships in the Imperial Russian Navy, high hopes were pinned on them. Four were picked to lead the Baltic Fleet on its voyage out to the Far East to make good the losses suffered at the hands of the Japanese. The ships selected were the *Borodino*, *Imperator Alexander III*, *Orel* and *Kniaz Suvorov*, and they led the motley collection of ships all the way to Japan. The fleet coaled at Nossi Bé in Madagascar, by courtesy of Russia's French allies. Port Arthur had fallen, leaving Vladivostok the only safe landfall. The ships ploughed on, their decks piled with bags of coal, and drawing far more water than the designers had envisaged for battle-conditions. As an example, in March 1905 the *Orel* displaced 16,800 tons with 2450 tons of coal on board; her draught was 32ft 6in forward and 30ft 6in aft.

The Battle of Tsushima on 27 May 1905 made history as a David vs Goliath encounter, although the discipline and training of the Japanese fleet levelled the odds considerably. The Russian formation's coherence was quickly destroyed by the superior tactics of the Japanese. This made Tsushima the first Nelsonic 'battle of annihilation' since Trafalgar a century earlier, but it is clear that if the Russians had been more disciplined and better led it might not have been such a walk-over.

The *Borodino* blew up when fires reached her magazines, the *Imperator Alexander III* capsized after heavy flooding through a large shell hole near the bow, the *Orel* was badly damaged by shellfire, and was eventually forced to surrender next day. The badly battered fleet flagship *Kniaz Suvorov* sank after hits by torpedoes (one 18in and between two and four 14in).

The *Orel* was rebuilt by the Japanese at Uraga in 1907 and was taken into service as the *Iwami*; she served until 1922, and was scrapped two years later. Significantly, the Japanese regarded her as carrying too much topweight, and the superstructure was cut down. The surviving unit, the *Slava*, had an active career in the Baltic during the First World War. She played a major role in the defence of the Gulf of Riga in 1917, and with her gun mountings altered to increase the elevation to 30 degrees, she outranged the German 12in L/50 guns in the dreadnought *König* (only elevating to 16 degrees). But on 17 October she sustained heavy damage from *König*'s guns in the Moon Sound. Trapped by her deep draught, she had to be torpedoed by Russian destroyers.

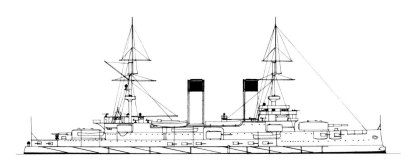

The Borodinos had thicker funnels and a lower conning tower than the Tsessarevitch*, and also a different armour scheme. (Line drawing 1/1250 scale.)*

THE BORODINO CLASS

Borodino laid down July 1899, launched 8 September 1901, completed August 1904 by New Admiralty Yard, St Petersburg

Imperator Alexander III laid down July 1899, launched 3 August 1903, completed November 1903 by Baltic Works, St Petersburg

Orel laid down March 1900, launched 19 July 1902, completed October 1904 by Galernii Island yard, St Petersburg

Kniaz Suvorov laid down July 1901, launched 25 September 1902, completed September 1904 by Baltic Works

Slava laid down October 1902, launched 29 August 1903, completed June 1905 by Baltic Works

Displacement: 13,516 tons (designed)
Dimensions: 397ft (oa) x 76ft 2in x 26ft 2in (max.)
Machinery: 2-shaft vertical triple expansion, 16,300 ihp; 20 Belleville boilers
Speed: 17.5–17.8kts
Armament: 4-30.5cm 40-cal (2 x 2) guns, 2-15cm 45-cal (6 x 2) guns, 20-7.5cm

	(20 x 1) guns, 20-37mm (20 x 1) guns, 4-38cm torpedo tubes (1 bow, 1 stern, 2 underwater)
Armour:	7.5-6in belt, 5.75-4in belt ends, 10-4in turrets, 6in secondary turrets, 3in 7.5cm battery, 8in CT
Fuel:	800-1520 tons coal
Complement:	835 officers and men

Conclusion

The design of the *Borodino* class emphasised all the regressive tendencies of late-19th-century battleship design, with too many light guns of dubious value, leading to too much weight devoted to protecting them. This was not a purely Russian vice; all navies were obsessed with close-range action, at ranges little greater than those in fashion in the days of sail. To be fair, this was not, as some historians have tried to claim, proof of admirals' resistance to change. Not until improved propellants became available in the last decade of the 19th century was it possible to fire at greater range, and not until the steel industry produced bigger forgings could long-barrelled guns be manufactured, but there remained the problem of firing accurately at long range. Unequal burn-rates of older propellants produced a random fall of shot, so battle-ranges had to be limited to distances at which individual gunners could spot their own shell-splashes.

Royal Navy analysis of the Russo-Japanese War actions, and in particular the Battle of Tsushima, were based on comments by British observers and by free access to the Japanese and ex-Russian ships repaired after the war. The Imperial Japanese Navy was not keen to publicise its experience and the lessons learned, but in view of the massive technical support given by the Royal Navy and British naval equipment suppliers, they had little option.

The first lesson was that a speed-margin was decisive. At Tsushima the Japanese enjoyed a substantial advantage for the first time; one observer estimated the average speed of the Baltic Fleet as no more than 9kts. The old ships were in poor condition and the long voyage had resulted in serious fouling of the hulls of all ships, leading to a loss of as much as 3kts.

Although the Japanese achieved prodigious rates of fire at Tsushima, and opened fire at the then stupendous range of 7000 yards, hardly any hits were scored at long range. Only when the range closed or the rate of change of range was low did the number of hits become significant; in the final stages the demoralised Russian ships were milling around. Both sides used primitive fire control instruments, 4ft 6in Bar & Stroud rangefinders in the Japanese ships and Krilov telescopic sights in the latest Russian ships. Barr & Stroud rangefinders were also fitted in the newer Russian ships, but during a trial off Madagascar the ranging errors varied between 7300 and 11,000m, possibly because no corrections were applied.

The Russian guns suffered from poor shell-fuses, and at Tsushima 8 out of 24 of the 12in hits on Japanese ships failed to detonate, as did 28 out of 81 smaller projectiles. This was, however, offset by problems suffered by the Japanese with their Shimose picric acid shell filling. This proved over-sensitive, in some cases causing premature detonation in the gun barrel, and bursting on impact rather than penetrating the armour.

Because the Russian ships were so overloaded they suffered very few hits on their belt armour. No penetrating hits were recorded by Russian survivors, either from ships which were not sunk or from survivors of the sunken ships. It seems unlikely that any Japanese shell penetrated Russian armour of 6in thickness or more, although the loss of so many ships at Tsushima makes the evidence less complete than the figures for the action on 10 August 1904. There were no penetrations of the *Orel*'s armour belt, although one 12in shell burst on 5.75in armour at the forward end. An incomplete Russian account claimed that the belt of the *Kniaz Suvorov* was not penetrated. As the *Orel* was captured the record of her damage is much more complete. The account fails to mention any penetration, whereas it was clear that every shell burst on impact.

There were problems with the supports behind armour. On 10 August 1904, the *Tsessarevich* had a 10in plate dislodged, resulting in flooding. The *Orel* had five out of eight fastenings broken on one plate. She was probably hit by five 12in shells, two 10in, nine 8in, and 39 smaller shells, on both sides. The hits were as follows:

The heavily damaged Orel *after the Battle of Tsushima. She was rebuilt by the Japanese and added to their fleet.*

Calibre	No.	Shell weight	Burster weights
12in	5	4200lbs	405lbs
10in	2	980lbs	96lbs
8in	9	2250lbs	207lbs
6in	39	3400lbs	351lbs

The weakness of conning-tower design has already been mentioned. At Tsushima personnel in the *Orel's* conning tower were injured by splinters driving in through the vision slits. Early in the battle the flagship *Kniaz Suvorov* was hit on her conning tower, causing heavy casualties and wounding Admiral Rojestvensky severely. Clearly the protection they offered was illusory, and the narrow vision slits restricted the view. It is significant that in the First World War the Royal Navy made little use of the heavy conning towers fitted in capital ships, and abandoned them post-war.

At Tsushima most of the Russian battleships were disabled by fires long before they reached a sinking state. Ammunition fires were common, but only the *Borodino* was sunk by a magazine explosion. Survivors' accounts describe graphically the effect of numerous fires caused by small-calibre shells. Firefighting parties suffered casualties and hoses were cut, allowing the fires to spread and join to form one massive conflagration.

From Russian accounts we can discern a pattern of common factors. These included a gradual breakdown in command as injuries or death eliminated senior officers. Voice pipes were cut, and access was frequently obstructed by debris, complicating the process of passing orders.

Movement on the upper deck was very dangerous because of the hail of splinters. Splinters also inhibited the stopping of holes above the waterline, and at times made repairs impossible. These holes led to an accumulation of water above the protective deck as the ships rolled in the heavy seas in the Tsushima Strait, and water from fire hoses exacerbated the problem.

The centre of gravity was high in the French-influenced *Borodinos*, with their lofty upper works; a satisfactory intact metacentric height was ensured by increasing the waterline beam. But much of the benefit of beam is lost when extensive flooding occurs, and we can be sure that their stability after damage was poor. The centreline bulkhead in the machinery spaces would lead to large heeling moments, while the righting moment would be seriously reduced if hits had made the upperworks non-watertight. The pronounced tumblehome would reduce the righting moment even further. The ships capsized from a combination of a high centre of gravity, asymmetric flooding and a reduced righting moment, although in the *Imperator Alexander III* flooding of the thinly armoured ends contributed to her loss of stability.

The sorry tale of the *Borodino* class should have taught very useful lessons, but already the major navies were moving ahead to radically new concepts such as HMS *Dreadnought* and the 'all-or-nothing' scheme of protection. The Russo-Japanese War had less effect because the advanced navies regarded even the Japanese as fighting with obsolescent ships and tactics, while the annihilation of the Russian fleet was widely attributed to sloth and incompetence. In an important sense, Tsushima was a live exercise between British technology and training on one side against French technology on the other, and there was little doubt which was superior. In another sense, Japan's astonishing victory over a major European power boosted its already inflated sense of superiority over the 'decadent West', and set the nation on the course of self-destruction which ended with abject surrender in Tokyo Bay in August 1945.

THE WORLD'S WORST WARSHIPS

DESTROYER HMS *SWIFT*

ROYAL NAVY 1906–1920

Admiral Sir John Fisher was a very human mixture of enthusiastic visionary and egoist whose legacy to the Royal Navy included HMS Swift.

The long-awaited appointment of Admiral Sir John Fisher as First Sea Lord in 1904 promised great things, and his adherents were not disappointed. He ushered in an era of rapid and sweeping reforms, many sound but others of more dubious value. On the plus side, he ensured political support for the highly innovative all-big-gun battleship HMS *Dreadnought,* pushed through much-needed reforms which improved conditions for the 'lower deck', and forced through a major strategic realignment by strengthening the Fleet in home waters. It is probably too much to expect Fisher, an extreme egotist, to have admitted his own shortcomings, but he was fascinated by technology without more than a superficial understanding of it.

The process of design and procurement of the big destroyer *Swift* is an excellent illustration of this contradiction in Fisher's character. Late in October 1904 he proposed to the Controller the construction of a much larger destroyer than anything built anywhere in the world. The intention was to create a new category, a vessel capable of fulfilling the usual functions of a destroyer, but seaworthy and fast enough to carry out scouting duties for the Fleet. Fisher saw such a super-destroyer as replacing the category of cruiser in fleet scouting. The Controller asked the DNC for his ideas on such a radically new design:

Displacement:	900 tons
Length:	320ft
Speed:	36 kts
Fuel:	Coal or oil
Weight of equipment:	77 tons
Weight of machinery:	435 tons

The DNC explained that the size of hull was totally inadequate, being only 20 tons more than the existing 550 tons of a *River*-class destroyer, and lacking the strength of older destroyers. In January 1905 a constructor was given just two hours to produce a legend, and he 'gave the best which could be done in the time':

56

Displacement:	1140 tons
Machinery:	19,000 shp
Speed:	33.5 kts
Hull weight:	450 tons
Machinery weight:	475 tons
Weight of equipment:	77 tons
Weight of armament:	33 tons
Fuel:	105 tons of coal

Fisher thought that speed was too low, so a second legend was prepared, with power increased to 29,000 shp and hull weight to 625 tons. In mid-April 1905 letters were sent to Armstrong, John Brown, Cammell Laird, Fairfield and Thornycroft, giving them only a month to submit proposals. This was an all but impossible deadline to meet, and all bidders asked for another two weeks, which was allowed. When the tenders were returned the cost was staggering, when compared to previous destroyers. For example, Armstrong produced the *Afridi* at a cost of £139,881, the *Cobra* at £63,500 *Swift*, and its tender for the *Swift* amounted to £284,000. John Brown managed to bring the cost down, but only to £191,717.

In June that year the DNC reported to the Controller that none of the designs was acceptable, and returned them for amendment. Not until mid-December could the DNC feel able to say that the revised Cammell Laird design met most of the requirements, in a destroyer 340ft long and displacing 1680 tons, subject to 54 changes. Within a few weeks the Admiralty decided that the 12pdr (3in) guns specified were to be replaced by 4in. John Laird was summoned to a meeting with the Board, at which a modified design was agreed. The armament was now agreed at four 4in guns and two 18in torpedo tubes, and endurance at full speed was reduced by two hours (from eight to six), and at economical speed. However, this design was also rejected, and a fourth version was drawn up, but this time the profile drawing was mislaid in the Admiralty, necessitating preparation of a copy, sent by the contractors in January 1906.

The element of farce is misleading; it really reflects the extremely tight deadlines imposed by Fisher. The Fisher style of hectoring and threatening retribution against anyone seen to be opposing him seems to have created an atmosphere of instant compliance, no matter how irrational the proposal might be. The DNC, Sir Philip Watts, was finally satisfied, and recommended the acceptance of the design at a cost of £236,000. The ship was to be oil-fired.

Despite the cost, the new ship, tentatively named *Flying Scud*, had a very light armament: two centreline tubes for launching 18in Mk VI* torpedoes, and four 4in Mk VIII guns, one pair sided on the forecastle, and two aft, mounted on the centreline forward of the mainmast and on a 'bandstand' platform on the stern. Early in April 1906 the name was changed from *Flying Scud* (a non-naval name) to the more traditional *Swift*.

The Board of Admiralty had optimistically included a premium of £18,000 for a knot more than the specified 36kts, but the penalties were severe. Late delivery would cost £400 per day, 0.25kts below the contract speed would cost £1500, 0.50kts £5000, 0.75kts £10,500, and 1kt £18,000. Work started in December 1906, and the ship was launched on 7 December 1907.

The preliminary trials were run over the Skelmorlie measured mile in March 1909, but the first run was abandoned when a boiler tube burst. The *Swift* could not reach her contract speed in the second trial, and new propellers were fitted, but on a displacement of 2135 tons she could barely get within a knot of the designed speed, reaching 35.099kts. After five hours a number of minor defects then developed, and the trial was stopped. What was particularly depressing was the fuel consumption, well over 27.5 tons per hour. In later trials no fewer than 26 different propellers were tried, in a vain attempt to meet the contract speed, but by September 1909 the speed was no better than it had been on the original preliminary trial six months earlier.

The Admiralty finally surrendered to the inevitable, and accepted the delinquent destroyer at a price of £236,764, without reaching her specified speed. In accordance with the contract, penalties totalled £18,000 for failing to reach 36kts, and an additional £26,240 for late delivery. The builders pleaded that the trials conditions had been unusually arduous, what with the frequent changes of propellers and the lack of room to manoeuvre on the Skelmorlie measured mile. As an 'act of grace' the Board agreed to limit the penalties to £5000, a tacit admission that the task imposed by the First Sea Lord was actually beyond any designer's capabilities. This failure did not, however, prevent Admiral Fisher from leaking spurious reports to the press, suggesting that the ship had reached 38kts. What is even more surprising is that authors of reference works still quote such preposterous figures today, to prove that Fisher had a genius for designing fast warships.

Initial reports of HMS *Swift*'s seakeeping were favourable. Her first commanding officer (CO), Captain Dumaresq, said that she more than justified her high cost when used for scouting. 'If all boilers are alight can jump from cruising at 10–12kts to 25 in seven or eight minutes…to 30 in twelve minutes, 34 in twenty minutes. Boilers can be kept disconnected, using scarcely any fuel, yet…be connected up and under full steam at from…10–15 minutes' notice…' Dumaresq also noted that she was vulnerable to damage from small shells, because large areas of her steam pipes were above the waterline. The Commodore (T), in charge of torpedo craft, was not so enthusiastic. 'For the small armament the cost is prodigious', and thought that she should have at least four torpedoes per tube, i.e. six reloads.

The enthusiastic reports from COs need to be put into context. The *Swift* was, of course, much bigger than any destroyer any of them had previously commanded or served in. Before the First World War torpedo craft operations in rough weather were generally curtailed when damage to structures was likely – the Admiralty did not like to get endless bills for repairs caused by over-enthusiastic young officers driving their ships too hard in heavy weather! War service would prove that the ship was too lightly built for service in Northern waters, however well she had performed in the southern North Sea.

Fisher's illusions about developing a new type of warship to bridge the gap between the battlecruiser and the destroyer, and make the light cruiser obsolete, came to nothing. His successor as First Sea Lord, Admiral Sir Arthur Wilson, minuted, 'I do not think we require any repetition of *Swift* in the immediate future'. The experiment of building super-destroyers was henceforth to be left to other navies, for even in what seems to us to be the free-spending pre-1914 Royal Navy, the Admiralty could not afford to build such expensive destroyers in the numbers needed. In 1912 a new design of light cruiser was begun, to make good the Royal Navy's serious shortage of scouting ships capable of working with the Fleet—the light cruisers that Fisher despised finally won the argument.

When war broke out on 4 August 1914, the *Swift* was assigned to the Grand Fleet, based at Scapa Flow in the Orkneys, as leader of the 4th Torpedo Boat Destroyer (TBD) Flotilla. This involved providing accommodation for the Captain (D) and his staff, increasing the ship's complement to 134. Modifications included the installation of high-speed anti-submarine sweep gear. Unfortunately she proved quite unable to cope with the winter weather, and had to be sent south to serve with the Dover Patrol late in 1915. This was a more demanding theatre, in the sense that the destroyers based at Dover were on constant call to deal with German raids on the mass of shipping ferrying stores and troops to France, but the waters of the Channel were relatively sheltered.

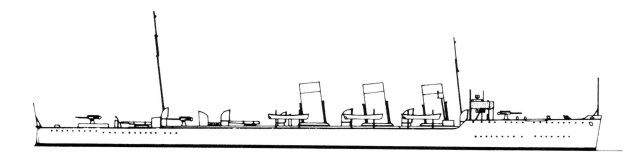

In 1916, to provide more firepower, the ship sacrificed her two forward 4in guns to make way for a 6in gun. This required stiffening of the decks below to compensate for the extra weight and the recoil. Like other attempts to over-gun destroyers it was not a success as it was hard to keep the gun trained on a target when the ship was rolling and pitching. In addition the flash when fired at night blinded bridge personnel (see below). Other improvements were added in this refit. The bridge structure was enlarged, with the navigating bridge extended out to the ship's side and the searchlight moved to a new upper bridge, both being protected by splinter-proof mattresses (of little use in keeping out shell fragments). The

HMS Swift *was a big ship, and no Royal Navy destroyer exceeded her displacement until the 1930s. (Line drawing 1/720 scale.)*

The radio room was moved from below to the charthouse on the starboard side of the new bridge structure. A sign of the changes happening in naval warfare was the provision of two anti-aircraft weapons aft, a 6pdr (57mm) gun and a 2pdr (40mm) pom-pom.

On the night of 20 April 1917 a force of six German destroyers from Zeebrugge mounted an attack on Calais and Dover. Two bombarded Calais and tried to rejoin their four flotilla-mates firing at Dover. Two flotilla leaders were patrolling off Dover, the ex-Chilean ship, HMS *Broke* (Commander Edward Evans) and HMS *Swift* (Commander Ambrose Peck). At about 11.30pm the sound of gunfire was heard from the direction of Dover and the two ships increased speed and headed in the direction of the gun-flashes. They found nothing after reaching the point where the flashes had been seen, and then returned to the east, arriving at a point about 7 miles east of Dover by about 00:50.

Suddenly a lookout spotted a line of six vessels steaming fast on a parallel but opposite course. They were identified as Germans because their coal-fired boilers were emitting sparks, and the *Swift* put her helm over to try ramming one of them. Both sides were firing, and Peck was temporarily blinded by the flash of the 6in gun, inadvertently passing astern of the enemy line. Either she or the *Broke* fired a torpedo which struck the fifth enemy destroyer, *G.85*. The *Broke*, seeing the explosion, altered course to ram the last ship, *G.42*, hitting her abreast of the after funnel. With both ships locked together there was terrible carnage on both sides, made worse by another German destroyer, which fired its guns at short range and caused even heavier casualties.

Meanwhile, the *Swift* pursued the fleeing survivors until, as predicted pre-war, she was forced to slow down after being hit several times. She rescued the survivors of *G.42* and gave assistance to the badly damaged *Broke* before returning to Dover. Both Peck and Evans were decorated and promoted.

It was the *Swift*'s moment of glory. After the action the 6in gun was removed and replaced by two sided 4in Mk V guns, but her career was drawing to a close. After the Armistice she was paid off and joined the long procession of over-age ships to the breakers, being sold for scrapping in 1921.

HMS *Swift*

Displacement:	2170 tons (normal), 2390 tons (load)
Dimensions:	345ft (pp) x 34ft 2in x 12ft 8in (deep)
Machinery:	4-shaft Parsons steam turbines, 30,000shp; 12 Laird boilers
Speed:	35kts
Armament:	4-4in BL Mk VIII (4x1) 2-18in torpedo-tubes for Mk VIx torpedoes
Complement:	126

Conclusions

The saga of HMS *Swift* shows what can go wrong when the technology is not sufficiently mature to match the mission envisaged. The *River*-class, with a speed of 25.5kts and a rugged hull, proved a very satisfactory improvement over their predecessors. But Fisher decided to give the next group, known as the 'Tribal' class, a massive increase of speed to 33kts, as well as oil fuel. The contractors were given only six weeks to produce a solution, and as a result no two were alike (a major drawback in time of war), and they suffered from very low endurance and only adequate seakeeping. The *Swift* took this dubious approach to destroyer design even further, resulting in a very expensive and disappointing outcome. True, she was an admirable effort to push the technology as far as it would go, and no Royal Navy destroyer exceeded her displacement until the late 1930s, but she was not a success.

The mission was also questionable. Although peacetime experience seemed to justify Fisher's hope that she could replace the cruiser as a fleet scout, she was so expensive that the type could never have been built in the numbers required. A return to more coherent designs showed that robust and well-armed destroyers were achievable—Royal Navy destroyers proved outstanding in battle, robust enough for service in all weathers, and yet cheap enough to be built in large numbers. By following a policy of progressive innovations they were able to advance to geared steam turbines (giving lower fuel-consumption) and heavier armaments. By 1918 destroyers half the size of the *Swift* were armed with 4.7in guns and six much more lethal 21in torpedoes

War experience was to show up the *Swift*'s limitations all too clearly, and she never functioned in the way that Fisher had intended, a common fate for extreme designs. That is hardly Fisher's fault; nobody in 1906 had much idea of how future naval warfare would turn out. But his dogmatic ideas on speed sacrificed that very flexibility which made the destroyer so successful as a type, without achieving sufficient additional benefits to justify the expense.

VIRIBUS UNITIS CLASS DREADNOUGHTS

AUSTRO-HUNGARIAN NAVY 1910–1918

One consequence of the the 1867 *Ausgleich* (Compromise), the equalling of the status of Austria and Hungary under the rule of their Habsburg monarch Franz Josef, was the creation of an Austro-Hungarian Navy in place of the old Austrian one. Elaborate efforts were made to iron out differences between the stronger Austrian economy and the less developed Hungarian economy. The Hungarian Delegation could block expenditure unless a share of the work was done on Hungarian soil, creating an administrative nightmare. One example of the pernicious results of this chaos had a major bearing on naval affairs: this was the insistence that part of the naval shipbuilding programmes was always allocated to a Hungarian shipyard. Hungary had in fact only 4km of coastline (including the port of Fiume, modern Rijeka), hardly conducive to an understanding of sea power.

The rapid heel of the Szent István *after being struck by two torpedoes fired from an Italian motor torpedo boat was captured for posterity by a cameraman on the next ship ahead.*

The Austrians (who until 1918 had extensive coastline on the eastern shore of the Adriatic) had a much clearer idea of the value of a navy, having vanquished the more powerful Italian Navy at the Battle of Lissa in 1866. Under the inspired leadership of Admiral Tegetthof the Austrian ships' aggressive tactics gave Austria the credit for promoting ramming from a dreaded consequence of poor ship-handling to a battle-winning tactic. Although Austria lost the war, the victory at Lissa led the Austro-Hungarian Government to pursue a goal of building a first-class navy to neutralise the Italian threat.

A series of handsome designs appeared, but they were generally smaller than their foreign contemporaries, and the relatively slow building-rates left the *Kaiserlich und Königlich* (KuK) Navy trailing behind. One contemporary commentator said that the greatest enemy faced by the KuK Navy was the government, both in Vienna and Budapest. The Dual Monarchy also suffered from the eternal clash of priorities faced by a land power: defence of the land frontier could never be neglected to spare money for naval expansion. Although Italy was the 'hereditary enemy', Russia's never-ending quest for a Mediterranean port also posed a threat.

The Navy had no lack of technically competent designers and engineers and a dedicated officer corps. In 1897 Admiral von Spaun became Commander of the Navy and Chief of the Navy Section of the KuK War Ministry, and tackled the problems as energetically as he could. His successor, Admiral Count Rudolf Montecuccoli, took office in 1904, and advanced the cause of the Navy by exploiting the support for it by the heir to the throne, Franz Ferdinand. One of Montecuccoli's finest achievements was to defeat a proposal to cut the naval budget by 50 per cent.

Less well-known was *Generalschiffbauingenieur* Siegfried Popper, the designer who was responsible for four classes of capital ships built between 1893 and 1911. His reputation among his professional contemporaries abroad stood very high, however, and he retired in 1907 loaded with honours. It is a tribute to his reputation that he was asked to return to duty to design the KuK Navy's first dreadnought battleships, the Viribus Unitis class (also known as the Tegetthoff class).

Popper's first sketch design was turned down by Montecuccoli on the grounds that it was too big and therefore unaffordable. The basic requirement was to build a counter to the first Italian dreadnought, the *Dante Alighieri*, armed with a dozen 30.5cm (12in) guns in four triple turrets, so Popper's design matched this. In fact, the cost-limits were to result in too cramped a design with too little margin of stability. With hindsight, it might have been better to reduce topweight by selecting an armament of triple turrets at weatherdeck level and twin superimposed turrets at shelter deck level.

All the major navies were at last taking the torpedo threat seriously, and Popper developed a theory, but without testing it on an old hull or a specially constructed target. His concept involved the fitting of armoured side sections in the double bottom, and the first ships so fitted were the Erzherzog Franz Ferdinand class pre-dreadnoughts built in 1907–11. He then tried a more extensive version in the Viribus Unitis class, 'cells' along the hull, armoured on the inside. It was similar to an idea tried in the small French battleship *Henri IV* and the French-designed Russian battleship *Tsessarevitch*.

Popper's design was deficient in internal subdivision. He did not believe in the anti-torpedo bulkheads favoured by the Imperial German Navy and the Royal Navy, among others. Instead, he clung to the outdated concept of a centreline bulkhead or did without any longitudinal bulkheads. War experience was to show that centreline bulkheads accelerated capsizing after flooding.

To add to all the technical problems, the Hungarian Delegation refused to vote any credits unless one of the class of four was built in a Hungarian shipyard. Unfortunately the only yard was the Danubius yard at Fiume, suitable only for the construction of small craft and needing major expansion and modernisation to build a dreadnought. A better site existed not far down the coast, but it could not be used for political reasons; it was not on Hungarian soil. Either site required heavy expenditure, but the delegations squabbled and wasted valuable time, and as a result the new ship, named *Szent István* after Hungary's patron saint, St Stephen, was seriously delayed. There is also evidence that the workforce's inexperience contributed to the loss of the ship (see below).

To get the other three started Count Montecuccoli had to risk his political career. Stabilimento Tecnico Triestino was short of work, and was about to close down part of its facilities and lay off skilled workers. The Commander of the Navy argued the case for placing two battleship orders urgently to the government in Vienna, but, as usual, the delegations argued and deferred a decision. Finally the Count took the extreme step of ordering two ships on his personal authority. Fortunately, the delegations and respective parliaments ratified his decision later. It must not be forgotten that the delinquency and irresponsibility of the politicians was taking place against a background of increasing international tension, and the Dual Monarchy was under threat from the Italians and Russians.

The main armament was supplied by the Skoda Works at Pilsen, and was excellent. The 30.5cm L/45 gun had a mounting that permitted 20 degrees elevation and a range of 25km (27,800 yds). The shell weighed 450kg. The choice of a triple mounting was influenced by the Italians' choice of a similar mounting for the *Dante Alighieri*, but the Italian ship was a long time building, so the Viribus Unitis class were the first in service with a triple mounting. The 15cm guns and the remainder were also supplied by Skoda.

At the outbreak of war the ships were allocated to Division I, but there was virtually no

On the surface, the Viribis Unitis class were handsome vessels, but they were outclassed by their contemporary rivals and had a disastrous combination of low stability and poor underwater protection.

action for them to show their fighting qualities as the Italian main fleet was distinctly reluctant to venture out of its main base at Taranto. Not even the reinforcements of British and French units could instill any wish to take risks. When in 1916 the Italian Government asked the British to move four pre-dreadnoughts to Taranto to 'guard against a breakout' the War Cabinet expressed its contempt for its allies in forceful terms.

As a 'fleet in being' the KuK battlefleet ran the risk of deteriorating morale, as happened with the Russians and the Germans. The siphoning off of the more energetic officers and ratings to torpedo craft and U-boats exacerbated the problem. In 1918 the new fleet commander, Admiral Horthy, was determined to strike at the enemy in the Adriatic, and planned a raid on the Otranto Barrage. This was a patrol line of drifters supported by Allied cruisers and destroyers, intended to prevent Austro-Hungarian and German U-boats reaching the Mediterranean from their base at Cattaro (modern Kotor).

A first raid on the Otranto Barrage by a force of destroyers had been only a partial success, and the Allies had repaired the damage quickly. Horthy's June 1918 raid was to be much more disruptive, and he planned to use his battleships to sweep aside resistance. Unfortunately the Italians, without knowing about the raid, had decided to station a small force of its MAS boats (motor torpedo boats) off Premuda Island in the southern Adriatic. Even that might not have proved fatal, but the *Szent István* was delayed by an overheated shaft, and one of the MAS was also delayed. On 10 June *MAS-15*, under the command of Lieutenant Luigi Rizzo, attacked the battleship in daylight, hitting the *Szent István* with two 45cm torpedoes.

Although the ship was hit in a boiler room the small warheads of the torpedoes should not have caused her to sink, but to everybody's amazement she started to heel rapidly. Later the huge hull capsized and sank quickly. The incident was recorded for posterity by a cameraman looking aft from the next ahead, the triple turrets dipping the mizzles of their guns in the water and the final capsize, with antlike figures scrambling over the upturned keel. The film-clip is renowned world-wide because a post-war charity looking after naval orphans sold copies to raise money. As a result almost every film-library in the world has a copy, but it also masquerades as footage from the Battle of Jutland, the scuttling of the High Seas Fleet, the sinking of HMS *Hood* and so on. It remains the only instance of a small strike craft sinking a capital ship at sea, although this scenario has remained the *raison d'etre* for small combatants ever since.

The rapid heel indicated a loss of initial stability, caused by an insufficient margin of stability. But much more serious was the collapse of bulkheads, caused by rivets popping under pressure; a second boiler room was flooded and a magazine. Even so, the ship took three-and-a-half hours to sink, because of heroic efforts by damage control parties pumping, moving weights and counterflooding. Mercifully the loss of life was relatively small, 89 sailors trapped between decks or drowned.

The official enquiry into the loss of the *Szent István* censured the commanding officer, several of the design staff and in particular Popper and the current *Generalschiffbauingenieur*, Fritz Pitzinger. Popper was told that the 'stiffeners' of his torpedo bulkheads were too weak. A notable absence from the enquiry was any constructor or engineer, who might have told the galaxy of senior officers that an adequate system of underwater protection was not compatible with restricted dimensions. They might also have recalled that money was not available for proper testing of target structures, i.e. firing live shells at armour plate, for which theoretical studies can never substitute. Nor did anyone point out that Popper's initial proposal was for a larger design.

In response to suspicions about the competence of the Danubius Shipyard's work-force, all technicians involved with the building of the battleship testified that they had tested all watertight joints, but had not subjected bulkheads to pressure-testing. It has also been suggested that lack of experience in the construction of large ships had led to faulty riveting. Incorrect heating of large rivets may set up metal fatigue, as can too much hammering.

The outcome was the wrecking of three distinguished careers. The captain had a good war record, but was declared unfit for further command. The widely respected Popper and Pitzinger were accused of incompetence and laziness. Nobody blamed the flawed system under which both laboured so long and loyally, and nobody raised the link between inadequate funding and limited dimensions.

Seen against the background of the crumbling Dual Monarchy, the affair was of little importance. On 31 October 1918 the *Viribus Unitis*, once the pride of the KuK Navy, was one of a number of ships handed over to the Commander-in-Chief of the new Yugoslav Navy, the Croatian Captain Janko Vukovic de Podkapelski. Early next morning two Italian combat swimmers entered the unguarded anchorage at Pola (now Pula) and placed a limpet mine on the hull. Tradition has it that the crew was celebrating the end of the war when the mine detonated, sinking her with heavy loss of life, including that of the captain. The motive for the Italian attack is generally agreed to have been a determination not to allow the Italian Navy to be robbed of its 'victory', and to frustrate any attempt to put the flagship beyond their reach.

The two survivors were given to France (*Prinz Eugen*) and Italy (*Tegetthoff*) as reparations. The French used their prize as a target, but Italy toyed with the idea of incorporating hers in its Navy for a while.

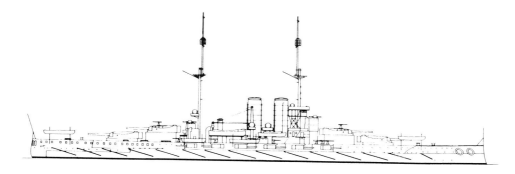

VIRIBUS UNITIS CLASS

The Viribis Unitis class was intended to match the Italian dreadnought Dante Aligheri, *but its design was constricted by lack of funds. (Line drawing 1/1250 scale.)*

Viribus Unitis laid down 24 July 1910, launched 24 June 1911, commissioned December 1912, built by Stabilimento Tecnico Triestino
Tegetthoff laid down 24 September 1910, launched 21 March 1912, commissioned 21 July 1913, built by Stabilimento Tecnico Triestino
Prinz Eugen laid down 16 January 1912, launched 30 November 1912, commissioned 17 July 1914, built by Stabilimento Tecnico Triestino
Szent István laid down 29 January 1912, launched 17 January 1914, commissioned 13 December 1915, built by Danubius, Fiume

Displacement:	20,000 tons (standard), 21,600 tons (full load)
Dimensions:	499ft 3in (oa), 495ft 3in (wl) x 89ft 8in (max.) x 29ft (deep load).
Machinery:	4-shaft Parsons steam turbines, 25,000 shp; 12 Yarrow boilers; *Szent István* two-shaft AEG steam turbines

Speed:	20kts (designed), 21kts (trials)
Armament:	12-30.5cm L/45 guns (4 x 3), 12-15cm L/50 guns (12 x 1), 18-6.6cm /50 guns (18 x 1), 2-4.7cm L/44 QF guns (2 x 1), 3-8mm machine guns, 4-45cm torpedo tubes (1 bow, 2 beam, 1 stern); 14 torpedoes
Range:	4200 nm @ 10kts
Fuel:	approximately 5600 tons coal + 162 tons oil
Complement:	993 officers and ratings +53 staff

Conclusion

Although the Viribus Unitis class marked a new high-water mark for the KuK Navy as the first dreadnoughts, and the main armament was very impressive, they did not compare well with their contemporaries. By 1911 the 30.5cm gun was outclassed by the new-generation 13.5in gun in the Royal Navy and its equivalent, the 14in in the US and Japanese navies. The Imperial German Navy may have influenced this, clinging as it did to the 28cm gun for too long, and switching to 30.5cm calibre when its rivals were moving up the scale. The appeal of the lighter calibres was theoretical; a lighter shell with a high muzzle velocity could achieve a greater range than a heavier shell. But in practice, a heavy shell with a low muzzle velocity kept its accuracy over the whole range, and reduced barrel wear.

The restricted dimensions limited speed, but the worst effect was on underwater protection. The combination of restricted beam and heavy armament reduced the margin of stability, even in an undamaged state, but underwater damage proved catastrophic.

All the KuK capital ships of the period suffered from the worst type of political interference:, resulting in inadequate funding and time wasted during the procurement process. In the long run, the Dual Monarchy was doomed to collapse under the weight of its inherent contradictions, but the story of the Viribus Unitis class is a tragicomic footnote to the naval arms race which played such an important part in the approach to the outbreak of war in 1914.

NORMANDIE CLASS DREADNOUGHTS

FRENCH NAVY 1913–1967

To a certain extent the choice of the *Marine Nationale*'s last pre-1914 battleships, the four *Normandie* class, breaks the rules of this study, for only one was completed, and converted to an aircraft carrier. But they embody so much of what was wrong and what was right about the pre-1914 generation of French warships that it is too tempting to exclude them.

The origin of the design can be traced back to December 1911, when the French Navy's Technical Committee examined the designs of the battleships to be laid down in 1912. These were the *Bretagne* class, the first to have their main armament of ten 340mm guns arranged on the centreline in five twin turrets. The Committee expressed regret at the failure of the Naval Staff to consult it at the planning stage. Midships axial turrets had been tried in the *Formidable* and *Amiral Baudin* in the mid-1880s and had caused blast damage to the superstructure, causing the Technical Committee to minute that this layout should be avoided if it proved possible to arrive at an acceptable disposition of multiple turrets 'which allowed the same number of 340mm guns to be mounted without any increase in displacement'.

France's director of naval contruction inferred from these comments that, if the Naval Supreme Council and the Minister of Marine were of the same opinion, the ships of the 1913 programme must be radically different. The first draft design was submitted in February 1912. The dimensions were driven by the depth of water in French harbours and capacity of dry docks: 170–172m length, 27.8m beam and about 8.8m draught. These figures resulted in a 'normal' displacement of 25,000 tonnes and a sped of 20–21kts, depending on the scale of armament. The Constructors wanted a speed of 21kts, with the same armament and protection as the *Bretagne* design, but also offered an alternative 20kt design armed with four quadruple 16in gun turrets.

Quadruple turrets had been proposed for the *Bretagne*, but the conventional twin turret was preferred. The design was drawn up by the St Chamond Company, with the guns mounted in twin cradles. Normally these pairs of guns would be loaded and fired together, but provision was made for independent loading and firing if one of the guns was damaged. The gunhouse was subdivided by a longitudinal 40mm screen, isolating the two sets of guns.

The Technical Department proposed two very different types of machinery: four-shaft direct-drive steam turbines (as in the *Bretagne*) and a hybrid system with direct-drive turbines on the two inner shafts and low-speed reciprocating steam engines on the outer shafts. The turbine was regarded as consuming too much fuel at normal peacetime cruising speeds (up to

16kts); the experience with first-generation turbines in the *Danton* class had not been satisfactory. However, late in the day it was decided to give the *Béarn* an all-turbine plant, with the original direct-drive turbines driving the outer shafts and using their exhaust steam to drive two low-pressure turbines on the inner shafts. To improve fuel-consumption cruising turbines were coupled to the centre shafts.

In March 1912 the General Staff declared in favour of retaining the 340mm gun, dismissed a St Chamond proposal for a triple turret, and accepted the quadruple turret. The preferred layout was two twin and two quadruple turrets, but if this was found to overload the ends too much, a *Bretagne* layout was acceptable. The draught was not to exceed 9m under any circumstances in full load condition. The Naval Supreme Council in April 1912 endorsed the 340mm gun and a layout á la *Bretagne*, unless a quadruple turret could be produced in time. The hybrid propulsion system was agreed, but not a new armour scheme. The Technical Department then began work on two designs, A7 (a *Bretagne* layout of five twin turrets) and A7bis (with three quadruple turrets). The latter arrangement saved some 240 tons on armament, and in total some 500 tons. The St Chamond proposals for the quadruple 340mm turret were accepted by the Minister of Marine on 6 April and a contract was signed in July. The original intention had been to arm the ships with a mix of 100mm secondary guns and 138.6mm guns, but the new 100mm mounting could not be ready in time, so a uniform secondary armament of 138.6mm was selected instead.

All the plans came to nothing when war broke out in 1914. Everybody 'knew' it would be a short war, and so the French Government (like its counterpart across the Channel) had no contingency plans. With no system of reserved occupations for key industrial workers, men in the dockyards and private shipyards downed tools when they were called up for military service. The heavy losses of soldiers and equipment, and the loss to the Germans of France's industrial heartland in the North led inevitably to the factories being turned over to land artillery and munitions. Work was stopped on all five battleships, and between September 1914 and May 1915 the hulls of the *Gascogne*, *Normandie*, *Flandre* and *Languedoc* were launched. In July 1915 an order ruled that completion of the ships was not a priority, and no new sub-contracts were to be placed. Later that month all work on their armament was stopped, but the Army needed the weapons, and so work continued on the production of 340mm guns for railway mountings. The 138.6mm guns were also pressed into land service, and when they wore out, were relined as 145mm artillery. A final order in January 1918 reiterated the suspension 'until further notice', but stipulated that stocks of material collected for them were not to be released without approval from the highest level.

In December 1917 the Constructors were asked to provide a summary of the percentage of work completed on each ship:

Ship	Hull	Engines	Boilers	Moving part of turrets
Normandie	65%	70%	delivered but used for destroyers	40%
Languedoc	49%	73%	96%	26%
Flandre	65%	60%	delivered but used for destroyers	51%
Gascogne	60%	44%	turbines delivered but used for destroyers	75%
Béarn	8–10%	25%	17%	20%

Only eleven days after the Armistice in November 1918 the Constructors sent the General Staff an outline of a project for modifying the Normandie class. The General Staff replied with a demand for speed to be increased to 26–28kts, better protection and more powerful armament. The Constructors replied by pointing out the difficulties and even impossibility of matching these requirements without a massive increase in expenditure and a very lengthy period needed for the work. There is no better example of the saying that the role of the Constructor is to tell the Naval Staff what it cannot have. The report reminded the Staff that the limitations on hull-dimensions had not changed since 1913; the enlargement of dockyards and drydocks had been seriously delayed. Only Brest had a 250m x 36m dock, while two at Toulon would not be ready for another year and those at Lorient and Bizerte would take even longer. Until then the beam of any new capital ships could not exceed 29.5m, which would limit the effectiveness of the anti-torpedo 'bulges' under consideration for the Normandies.

The General Staff recognised that any new design reflecting war-experience would take at least six or seven years, and Admiral de Bon wrote in February 1919, 'I think we must resign ourselves to completing the four *Normandies*.' He set out the Staff's general conclusions:

1. For the four ships already launched, retain the main machinery. Raising speed to 24kts would require 80,000 shp (from the existing 32,000 shp) and would be an expensive and lengthy process.

2. Improve underwater protection by filling outer compartments with cork chips, except for those already containing coal, and fit a 1-metre wide 'bulge'. The latter would provide a greater margin of stability to compensate for weights added elsewhere, and would increase resistance to a torpedo-hit from 100kg in the 1913 design to 200kg.

3. Increase the elevation of the 340mm guns to 23–24 degrees to increase range to 25,000m. Transfer the *Béarn's* turrets to the *Gascogne* to replace those captured by the Germans at Fives-Lille. The General Staff also hoped to get rid of the midships quadruple turret and revert to the 1912–13 idea of twin superimposed turrets forward and aft.

4. Increase deck protection to 120mm and ensure that any diving shell would have to penetrate three decks before it could reach the vitals.

5. Replace underwater 450mm torpedo tubes with six above-water 550mm fast-loading tubes.

6. Improvements to fire control for main and secondary armament and torpedo tubes.

7. Make provision for a two-seater spotting aircraft.

For the moment the question of completing the *Béarn* was in abeyance. She was launched in April 1920 to clear the building slip while the debate on the future of the four sisters continued. The Technical Department was investigating the possibility of lengthening their hulls by 15m to increase speed by as much as 5kts. But when Admiral Ronarc'h succeeded Admiral de Bon as Chief of the General Staff he tried to bring a new urgency to the discussions. In July 1919 he pointed out that the Italian Navy was the obvious rival, and that work might restart on the four 29,000-ton *Carraciolo* class, suspended since 1916. Ronarc'h listed three possibilities:

1. Complete the ships to the original plans, but at higher cost on account of inflation.
2. Improve range of guns and strengthen protection.
3. Lengthen the hull and increasing power to achieve higher speed.

He also advocated cuts in the naval programme to reduce the burden on the French economy, which had come close to collapse during the war and was recovering very slowly. On 12

September 1919 he wrote that the Navy should give up the idea of completing the *Normandies*. This was to be the death sentence, although ardent supporters of French sea power continued to promote crackpot schemes. The formal cancellation was included in the preamble to the 1922 Programme, and four of the hulls were laid up in the 'graveyard' at Landevennec, stripped of useful material and scrapped in 1923–26.

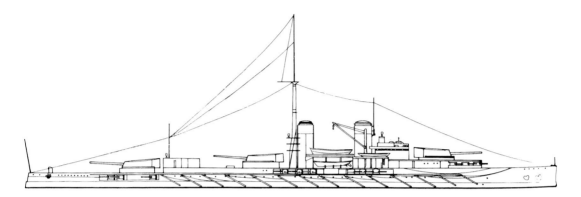

The Normandie class symbolised all that had gone wrong with French naval policy at the turn of the century and in the decade that followed. (Line drawing 1/1250 scale.)

NORMANDIE CLASS

Normandie laid down 18 April 1913, suspended 1914–15, stricken 1922, built A & Ch de la Loire, St Nazaire
Languedoc laid down 18 April 1913, suspended 1914–15, stricken 1922, built Forges & Ch de la Gironde, Bordeaux
Flandre laid down 1 October 1913, suspended 1914–15, stricken 1922, built Brest Naval Dockyard
Gascogne laid down 1 October 1913, suspended 1914–15, stricken 1922, built Lorient Naval Dockyard
Béarn laid down January 1914, suspended 1914–15, launched April 1920, restarted as carrier (see below), built Forges & Ch de la Mediterranée, La Seyne

Displacement:	25,230 tons (standard)
Dimensions:	170.6m pp, 175.6m wl, 176.6m oa x 27m x 8.65m
Machinery:	2-shaft steam turbines/2-shaft reciprocating steam, 4000 shp/2-shaft reciprocating, ihp; 21–28 boilers
Speed:	21kts
Armament:	12-340mm 45-cal guns (3 x 4), 24-138.6mm 55-cal guns (24 x 1), 6 underwater 450mm torpedo tubes
Armour:	30cm side; 18cm casemates
Fuel:	2700 tons coal, 300 tons oil (maximum)
Range:	1800 nm @ full speed; 3375 nm @ 12kts
Complement:	43 officers, 120 POs and 1037 ratings

In 1922 it was finally decided to complete the *Béarn* as an aircraft carrier. The enormous advances in naval aviation, particularly in the Royal Navy, had largely passed the French Navy by, apart from a small number of Cross-channel steamers converted to seaplane carriers. The battleship hull was relatively simple to convert as it was not very advanced at the time that work had stopped in 1915. Even so, it stretched the French Navy's resources, and some technical assistance was given by the Royal Navy, which made available details of its own converted battleship, HMS *Eagle* (ex-*Almirante Cochrane*).

The powerplant was not the one planned for her in 1913, but the hybrid turbine/reciprocating engine installation intended for her four sisters. Conversion started at her original builders, Forges et Chantiers de la Mediterranée, at La Seyne in Toulon in August 1923. The work was completed in May 1927, giving the French Navy its first fleet carrier. The hangar had three electric lifts connecting the 124m-/407ft-long hangar to the 180m/590ft flight deck. In theory she could operate 40 aircraft, but the General Staff's insistence on a heavy anti-ship armament of eight 155mm guns in casemates encroached on internal volume and cost valuable weight. A pointless feature was the inclusion of four 550mm fixed underwater torpedo-tubes.

The *Béarn's* slow speed was a serious handicap, preventing her from exploiting advances in carrier aircraft performance. When the Second World War broke out she was used to ferry aircraft from Canada to France. When France surrendered in June 1940 she proceeded to Martinique in the West Indies, where she remained inactive until 1944, when she went to the United States for a full refit, rearmament and conversion to an aircraft transport. After the war she continued in her new role, ferrying aircraft to Indochina during the conflict with the Vietminh. After her retirement she served as an accommodation ship for submarine crews at Toulon, and was finally scrapped there from March 1967. The stripped hull was finally sold to Italian shipbreakers.

The Béarn *was France's only aircraft carrier in the Second World War, but was already obsolescent in 1939, a consequence of technological advances since the decision was taken to convert her in 1922.*

BÉARN

Displacement:	22,146 tons (standard), 25,000 tons (full load)
Dimensions:	559ft 7in (pp), 598ft 11in (oa) x 115ft 5in (maximum) x 30ft 6in
Machinery:	4-shaft, reciprocating steam (outer shafts), turbines (inner shafts); 12 Normand du Temple boilers; 15,000 ihp + 22,500 shp
Speed:	21.5kts
Protection:	83mm side; 25mm (flight and main decks); 70mm casemates
Armament:	8–155mm 55 cal (8 x 1), 8–37mm AA (8 x 1), 4–550mm torpedo tubes (underwater)
(After 1944):	4-5in 38 cal (4 x 1), 24-40mm AA (6 x 4), 26-20mm AA (26 x 1)
Aircraft:	40

Fuel/endurance:	2160 tons oil/6000 nm @ 10kts
Complement:	45 officers, 830 ratings (reduced to 27 officers + 624 ratings in 1944)

Conclusion

The long technical debate about the configuration of the *Normandie* design in 1912–13 and again in 1919–20 showed how far French ambitions outreached their capabilities. The organisation had never recovered fully from the depredations of the Minister of Marine Camille Pelletan in his short reign in the early years of the century. France also faced the more immediate threat of a land war with Germany, so adequate funding to modernise the Navy was never available. In fact the *Entente Cordiale* (1904) was a tacit recognition of this dilemma, and sought to remedy deficiencies by relying on the Royal Navy to defend the coast of Northern France from a German attack.

In addition to these strategic problems, France faced the problem of its weak industrial base. Type for type, French warships were no longer as good as their equivalents in other major navies, and efforts to catch up were not effective. Like other navies in long-term decline, its designers sought advanced solutions to redress the balance, relying on revolution rather than evolution. The French Navy had never been strongly influenced by foreign designs, secure in the knowledge that French technology was the best in the world. The agonised debate over the *Normandies* was evidence of a lack of a clear tactical objective. On the other hand, the adoption of quadruple turrets was nearly 25 years ahead of its time.

The *Béarn* was not the most successful carrier of her day. Speed was too low for efficient operation of aircraft, a weakness that was made worse by improvements in aircraft performance. The only navy to achieve successful conversions was the US Navy, with its large-hulled *Lexington* and *Saratoga*; smaller hulls were expensive on internal volume, resulting in a small air group. What is surprising about the *Béarn* is how long she lasted, although her active career as a carrier lasted only 12 years.

AA CLASS FLEET SUBMARINES

US NAVY 1914–1930

The US Navy's first fleet submarine, the USS *Schley* (SS-52) was authorised in 1914, followed by two unnamed sisters a year later. They entered service as *AA-1* (numbered on 24 August 1917), *AA-2* and *AA-3*, with new hull numbers SF-1, SF-2 and SF-3 respectively. The design agency was the Electric Boat Company; *AA-1* being built to design 63A, while her sisters were built to the slightly improved 63C design.

The design was unusual, a partial double hull and two single-hull compartments forward and aft, using riveted 0.5in steel. The two end compartments and the Central Operating Center (COC), forerunner of the control room, were designated 'refuge compartments' and fitted with double watertight doors to withstand pressure from either direction. A small conning tower above the COC was fitted with an escape trunk, and although evidence is sketchy, escape trunks were probably fitted in the other two 'refuge compartments'. Hatch-sizes were not standard; 22in for the conning tower, 36in for the engine room (to permit the installation and removal of machinery components) and 27in elsewhere.

The two pairs of trainable 18in torpedoes fitted in *AA-1* were aimed through the periscope, trained by large handwheels inside the hull, and fired through doors on either side of the superstructure. They were replaced by 21in torpedoes, but during trials in 1920 the above-water tubes were found to be inaccurate. One set was immediately removed, followed a short time later by the other pair. The designed gun armament was two 3in 23-cal guns on folding mountings, but soon after completion *AA-1* was rearmed with a 4in 50-cal 'wet' mounting, and the change was incorporated in *AA-2* and *AA-3* during construction.

All five engines were installed in a single compartment, with two in tandem on each propeller shaft, and an electric motor was mounted on each shaft. The electric motors were connected, allowing one to be used as a generator to charge the batteries. There were two banks of 60-cell batteries, which could each be connected to the motors, either in series or in parallel. The arrangement was chosen to control shaft-speed by varying the voltage. The tandem arrangement caused severe vibration, and a series of different clutches and dampers could not cure the problem.

The main ballast tanks were flooded by manually-operated Kingston valves, via linkages to the COC. The tank structure was in fact very complex and unusual, with some tanks on top of others or inside them. Vent valves at the top of the ballast tanks were operated by hand and the tanks were drained into a 'main drain' in the duct keel, which had high-pressure, low-pressure

The AA class were unsuccessful because they were not fast enough to keep up with the fleet, and were effectively the wrong kind of ship for the task they were required to perform.

and emergency ballast pumps. The only other auxiliary machinery was a transfer pump for diesel oil, steering and diving-gear motors, and a windlass for the two anchors. One of these anchors was housed topside and the other was below the keel in a recess. The bow diving plans were shaped like round flippers, and folded back against the hull when not in use. Apart from searchlights and flag-hoists, the only means of communication was a primitive radio transmitter and receiver, and a Fessenden oscillator for short-range signalling under water.

Although the 'AA' class had no difficulty in making their designed speed of 20kts (surfaced) and 10.5kts (submerged) under ideal conditions, seakeeping was not good. The low, narrow bow caused them to trim by the head and 'take it green' over the deck; the problem was not made any better by the low reserve of buoyancy. In an effort to cure this weakness the forward main ballast tank was given stronger plating to allow it to be a buoyant space, but it was ineffective.

The three boats were renumbered *T-1* to *T-3* in September 1920. In an effort to improve performance the T-3 was re-engined with two large German MAN diesels, but this proved ineffective. In fact there is some doubt about the origin of these diesels; they were reported to have been taken out of one of the U-boats handed over to the USN as war prizes, but it is equally likely that they were an American adaptation of an MAN design. Other improvements have already been mentioned, but there was little that could be done. They had taken too long to build—being completed in 1922—and the rapid advances in submarine warfare during the First World War rendered them obsolete. *T-3* was the first to be decommissioned, in November 1922, followed by *T-1* a month later and *T-2* in July 1923, probably the shortest active life of any US Navy submarine, and possibly in the world. They were declared redundant to comply with the London Naval Treaty, and stricken in September 1930.

AA CLASS

AA-1 laid down 21 June 1916, launched 25 July 1918, commissioned 30 January 1920
AA-2 laid down 31 May 1917, launched 6 September 1919, commissioned 7 January 1922
AA-3 laid down 21 May 1917, launched 24 May 1919, commissioned 7 December 1920
All built by Fore River Ship & Engine Co. (sub-contracted by Electric Boat)

Displacement:	1106 tons (surfaced), 1486 tons (submerged)
Dimensions:	268ft 9in (oa) x 22ft 4in (max.) x 14ft 1in
Machinery:	2-shaft 4000 bhp diesels; 2-1350 shp electric motors
Speed:	20kts (surfaced)/10.5kts (submerged)

Armament:	8-18in torpedo tubes (4 bow and 2 x 2 18in (deck) *T-1* rearmed with 21in torpedoes in 1920 and reduced to 4-21in (bow only); *T-1* 16 torpedoes, others 12 torpedoes *AA-1*: 2-3in 23 cal. guns (replaced by 1-4in 50cal. gun; *AA-2* and *AA-3* completed with 1-4in gun)
Diving depth:	150ft
Range:	3000 nm @ 14kts (surfaced)
Fuel:	approximately 24,000 gallons
Patrol endurance:	30 days
Complement:	4 officers + 5 CPOs + 45 enlisted personnel

Conclusion

These fleet submarines, the first in the US Navy, had many innovative features but they were unsuccessful for a variety of reasons. The *rationale* for them came from a vocal pre-war lobby that saw large submarines as replacements for destroyers screening the battle fleet. They bear superficial comparison with the Royal Navy's K class (see page 84), but lacked sufficient surface speed to operate with the Fleet. Nor did anyone seem to be aware of just how limited the visibility was from the conning tower; for the same reason, submarine gunnery was a very crude affair. Without even the crudest fire control, submarine gunfire was useful only at short range against very 'soft' targets. They were very complex, slow to dive and hard to handle.

By the time the AA boats came into service the US Navy had gained some limited experience operating in small numbers in the European theatre. But this was in highly specialised anti-submarine warfare, like their British allies; by 1918 any chance of a fleet action was so remote that they would have had nothing to do.

The US Navy did not lose interest in fleet tactics for submarines, largely because the limitations of the AA class were not yet known when nine more fleet types were authorised from 1921 onwards and laid down in 1921-34. The first three *V-1* class, more often known as the B class, were authorised in 1916 as *SF-4* to *SF-6*, but were renumbered SS-163 to SS-165 in 1920, when a standard hull-numbering system was introduced.

The first three, *V-1*, *V-2* and *V-3*, were given the names *Barracuda* (SS-163), *Bass* (SS-164) and *Bonita* (SS-165) in February 1931. The preliminary design work was done by the Navy's Bureau of Construction & Repair, and details were worked out by the builder, Portsmouth Navy Yard. They had a riveted partial double hull, subdivided into nine watertight compartments. The partial double hull featured a complex arrangement of tanks, and a non-circular cross-section that was different in every compartment. Other unusual features included a complex single-casting hatch structure containing both access for loading torpedoes and the escape trunk. A very unorthodox bow-shape was adopted, but it made the boat plough into the waves rather than riding them, despite the provision of two buoyancy tanks in the bow. The bridge was made taller than in previous designs, to improve visibility when running on the surface.

The designed speeds, 18.7kts surfaced and 9kts submerged, were more realistic than those of the AA class. Endurance was traded off against speed, resulting in 6000 nm (normal) and 10,000 nm (maximum at 11kts); submerged endurance was 10 hours at 5kts. The propulsion plant featured direct drive as well as diesel-electric drive; it was the precursor of all-electric drive, in which diesel generators provided current for the electric motors. However, both main diesels and electric motors were a constant source of trouble.

No tears were shed when the three boats were laid up in reserve in 1937. They were a great disappointment to those who supported the outdated concept of a fleet escort. Although full of interesting innovations, the design showed no appreciation of wartime experience. They were saved from the scrapheap by the Second World War, but they continued to earn a bad name for unreliability. A measure of desperation was the conversion to cargo-carriers in 1942–43, but they were equally unsuited to that task. They were decommissioned before the end of hostilities, unmourned and unloved.

The next variant was the one-off *V-4* (SM-1) which was conceived as a 2700-ton hybrid minelayer/cruiser submarine, capable of laying 60 Mk XI mines through stern-chutes, and armed with two 6in guns. Renamed *Argonaut* in February 1931 (her allocated hull number SS-166 was never official), she was never regarded as successful. Although this may have been in part reflected her operators' lack of familiarity with the complex minelaying gear, she was too underpowered, and never reached her designed speed of 15kts (13.65kts was all she could manage on trials). She was also too large to be used as a fleet submarine, lacking manoeuvrability.

The USS *Argonaut* was to receive new engines in 1941 but the outbreak of war in December 1941 caused the modernisation to be postponed. This was eventually done at Mare Island Navy Yard, however, but before she could leave Pearl Harbor for a long-range patrol she was hurriedly converted to a transport submarine (APS-1) capable of transporting 120 marines. She took part in the raid on the Makin Islands in 1942, with reasonable success, and was then sent to Australia. She suffered the misfortune to be sunk by Japanese escorts near New Britain on 10 January 1943, with the loss of 105 officers and men.

The next batch of so-called 'Bs', *V-5* and *V-6*, resembled *Argonaut* but without the minelaying capability, compensated for by two stern 21in torpedo tubes aft, in addition to the four bow tubes. Named *Narwhal* (SF-8) and *Nautilus* (SF-9), they were given new hull numbers SC-1 and SC-2 in 1925, and shortly after they were commissioned in 1930 they were given new names and numbers: *Narwhal* (SS-167) and *Nautilus* (SS-168). Like *V-4* they were mediocre, slow to dive, clumsy underwater, and easy to detect. Once again they proved unable to match their designed speed of 17.44kts, achieving a maximum of 14kts on trials. They were due to be re-engined in 1940–41 but the outbreak of war prevented the work being done, although *Nautilus* was fitted with tankage for 19,000 gallons of aviation fuel, in order to refuel seaplanes. She entered Mare Island Navy Yard in July 1941 for re-engining, and *Narwhal* had a similar overhaul a year later. Both were converted to transport submarines, and served until the end of the war in August 1945.

The *V-7* (SC-3) marked a retreat from giant submarines, and although still classified as a submarine cruiser, was closer to the later standard 'fleet' submarines which inflicted so much destruction on Japanese shipping. While still under construction she was renamed *Dolphin* (SS-169) and this time her designer, Andrew I McKee, proved that the US Navy would get better value from a smaller submarine, without sacrificing armament or habitability. She had no trouble in reaching hr designed surface speed of 17kts, and operated in the Pacific from 1933. Unfortunately she was never anything but a prototype, being one-of-a-kind with many non-standard items of equipment. For that reason she was regarded as not worth the cost of modernisation, and for most of 1943 was used at Pearl Harbor to train submarine crews. She was sent back to Mare Island for 'patch repairs' and passed on to Portsmouth Navy Yard. She continued to serve as a training boat at the New London submarine base with a depth-restriction of 150ft, for the rest of the war.

The *V-8* class marked a final retreat from the old-fashioned 'fleet' or cruiser submarines so

popular after the First World War. In fact the US Navy was only one of a number of navies which were victims of the inflated reputation of the German '*U-kreuzern*'. When the 'fleet' designation was resurrected by the US Navy it was to be used in the sense of long-range, high-performance and with a heavy torpedo-armament.

Although not an unqualified success the *V-8* and *V-9* (renamed *Cachalot* and *Cuttlefish* with hull numbers SS-170 and SS-171 in 1933) tested many items, such as air-conditioning, for the new generation of fleet boats which dominated the underwater war in the Pacific. It had been a long and hard road from the AA boats, but it meant that the US Navy entered the Second World War with submarines tailored to the needs of the Pacific War.

The USS Bonita *was one of three submarines that incorporated changes that offered slight improvements over the AA class, but she was of a complicated construction and not a good sea boat.*

'FLUSH DECKER' DESTROYERS

US NAVY 1914–1952

The US Navy's approach to destroyer tactics before the First World War was very different to its European counterparts. There was no equivalent of the light cruiser, apart from three scout cruisers (the Chester class, briefly discussed in the chapter on the Omaha class, page 102) and it was believed that the scouting and battlefleet escort functions could be undertaken by destroyers.

Although destroyers had been built in small numbers there was no continuity. The first modern destroyers were the *Smith* (DD-17) class, three of which were authorised in 1906, followed by two in 1907 (DD-20-21). On 8 March 1907, the day after Congressional approval was given, the Board on Construction approved the design. It was to be driven by 10,000ihp reciprocating or 9000shp Curtis turbines; speed at normal load was to be 28kts, 26kts at deep load. Known as the 'flivvers' (possibly a reference to the Model T Ford), they displaced 700 tons and were armed with three single 18in torpedo tubes and five 3in guns. An unusual feature was the provision of enclosed stowage for a single reload torpedo close to each tube.

The US Navy faced the same problem as other navies when changing to steam turbines for destroyers: these were efficient at high speed but inefficient at low speed. At high speed they developed a very high power-output; the revolutions per minute were about twice what is required for modern turbines. The importance the US Navy attached to the scouting duties of destroyers, made it essential to maximise endurance for a given expenditure of fuel. One solution adopted in some destroyers was the provision of reciprocating cruising engines.

More 'flivvers' were built, the 20 *Paulding* group (DD-22-42) group authorised in Fiscal Years 1908 (ten), 1909 (four) and 1910 (six). The main differences were an increase of power to 12,000 shp, raising the designed speed by 1.5kts, and a change to three twin 18in torpedo tubes. The General Board's campaign to increase the Navy's destroyer-strength was bearing fruit.

The greater power installed in this group provided higher speed and greater economy, and the adoption of oil fuel was another step in the right direction. The triple turbine powerplant was based on experience with the 'flivvers', but *Perkins* (DD-26) and *Sterett* (DD-27), built by Fore River Shipbuilding, were driven by only two Curtis turbines, and *Warrington* (DD-30) and *Mayrant* (DD-31), built by Cramp, had two Zoelly turbines. Most of the 20 units had four evenly spaced funnels, but the *Roe* (DD-24), *Terry* (DD-25), *Perkins*, *Sterett*, *Warrington*, *Mayrant*, *Monaghan* (DD-32), *Walke* (DD-34) and *Patterson* (DD-36) had three, the middle one thicker than the other two.

A broadside view of the USS Jacob Jones (DD-130), one of the Wickes class of flush-deck destroyers, in December 1917. She was sunk off the coast of New Jersey by a U-boat in 1942.

The General Board was anxious to push destroyer-displacement to 1000 tons because all its theoretical studies convinced its members that the most valid role for destroyers was the protection of the battle line. This implied that the number of destroyers must be determined by the number of battleships. The 1903 plan had allocated one destroyer to each battleship, but four years later this allocation had risen to four per battleship because of the increased threat posed by torpedo attack.

In 1909 the General Board was given responsibility for the characteristics of future warships, replacing the Board on Construction. In 1910 the new Board suggested duplicating the 'flivver' design, especially 'in view of the success obtained by the *Flusser* [*Smith*] class'. A June 1910 memorandum spelled it out: 'they must be able to keep up with the fleet under all probable conditions of weather and distant cruising...' It was hoped to achieve a radius of 4000 nautical miles at 15kts, and make 30kts for one hour or 25kts for 24 hours. The desired armament was a 4in or preferably a 5in gun to replace one 3in, but the 5in gun proved too ambitious. In the final design the Bureau of Construction & Repair replaced all the 3in guns with 4in, a major increase in gunpower. Between September 1910 and March the following year eight sketch designs were prepared, and Congress authorised eight new destroyers in March 1911. These were the *Cassin* class (DD-43-50), known in service as the 'thousand tonners'.

In assessing these rapid advances in design, it must not be forgotten that very few of the 'flivvers' had been completed, so there was little opportunity to evaluate the new features. In fact, defects came to light only after all of them had been authorised. The main drawback was the stern torpedo tube, which could not fire torpedoes accurately above a speed of 20kts because of the effect of the stern wave. Little could be done about the 'flivvers', but the Chief Constructor suggested that in the *Cassin* class the stern tube and the after 4in could be interchanged, and this was done.

The General Board was sufficiently happy with the 'thousand tonner' design to put it forward for the 1913 progamme. But the warring tribes among professional destroyer officers were not so happy; some thought the 'thousand tonners' were too big and others wanted a heavier torpedo-armament. The Naval War College supported an increase to the massive armament of four triple torpedo tubes in beam positions *and* reloads. The Board refused to countenance such an extreme position, and accepted a Bureau of Ordnance compromise: the addition of two torpedo tubes, a change to the much more powerful 21in torpedo, and eliminating one of the 4in guns. The Board incorporated the new scale of armament in the next group of 'thousand tonners', the *O'Briens* (DD-51-56).

More attention was being paid to machinery improvements. For the last group of *Cassins*, Cramp proposed a two-shaft turbine installation, with reciprocating engines for cruising. The change required an altered stern, and the *Aylwin* (DD-47) failed to make her designed speed on trials. One of the slightly improved *Tucker* class, the *Wadsworth* (DD-60) was the first US Navy destroyer with main turbines geared to the propeller shafts. Unfortunately, she only made only 30.7kts on a trials displacement of 1050 tons in July 1915, but 326 tons of oil fuel drove her for 5640 nautical miles at 16kts. The *Cassin* (DD-43), *Cummings* (DD-44), *McDougal* (DD-54) and *Ericsson* (DD-56) each had a reciprocating engine that could be clutched to one shaft for cruising below 15kts. The *Tucker* (DD-57), *Conyngham* (DD-58), *Porter* (DD-59), *Jacob Jones* (DD-61), *Wainwright* (DD-62), *Allen* (DD-66), *Wilkes* (DD-67) and *Shaw* (DD-68) differed in having a single cruising turbine geared to one shaft. The remainder had a more symmetrical arrangement, with two reciprocating engines for cruising in the *Aylwin*, *Duncan* (DD-46), *Parker* (DD-48), *Benham* (DD-49), *Balch* (DD-50), *O'Brien* (DD-51), *Nicholson* (DD-52), *Winslow* (DD-53) and twin cruising turbines in the *Cushing* (DD-55).

It is interesting to read of a meeting between Royal Navy officers of the cruiser HMS *Suffolk* and US Navy officers in Mexican waters in May 1914. The British officers were amazed to hear that US Navy destroyers regularly cruised at 20kts, whereas they themselves were limited to 15kts except during the annual manoeuvres. This was interpreted by the Bureau of Steam Engineering as the Royal Navy keeping its destroyers in good mechanical condition, whereas the prolonged hard driving of US Navy destroyers caused a large number of breakdowns.

The Board devoted considerable effort to the 1916 Destroyer, trying to balance the unrealistic scouting function against the 'normal' destroyer function of attacking and defending capital ships in a fleet battle. The Bureau of Construction & Repair suggested the radical innovation of a flush deck and triple torpedo-tubes. The flush deck was intended to strengthen the hull to improve seakeeping and enable the heavy armament to be carried; a conventional high forecastle would have resulted in a shallower hull girder and an unacceptably weak hull. Six destroyers, the *Caldwell* class (DD-69-74), authorised in March 1915, thus became the prototypes for the mass-production programme needed in 1917. Two, the Cramp-built *Conner* (DD-72) and *Stockton* (DD-73), had the three-shaft machinery planned by the Bureau of Construction & Repair. The *Caldwell* (DD-69), had General Electric-built Curtis turbines with separate cruising turbines linked to them by 'electric speed-reducing gear', whereas the *Craven* (DD-70), *Gwin* (DD-71) and *Manley* (DD-74) had Parsons geared turbines. The *Conner* and *Stockton* and the Seattle-built *Gwin* had three funnels, whereas the others had the standard four-funnel layout.

When the US Government joined the Allies as a 'co-belligerent' in April 1917, the endless arguments over destroyer characteristics became irrelevant. The Navy suddenly needed large numbers of ships, and in view of the crisis in the war against the German U-boats, the message from the Admiralty was 'send us all the destroyers you can'. The solution was to adapt the experimental *Caldwell* design, despite the fact that it had been intended as a fleet destroyer, and would have to operate as an anti-submarine vessel. The reason was the need to avoid disruption or the delay inherent in designing a specialised 'austere' escort.

Space does not permit a detailed analysis of the massive American destroyer-building effort in 1917–18, but it was a prodigious achievement. The 1916 programme authorised the building of 50 destroyers (DD-75 to -124), of which DD-75 to -94 were to be started immediately. Fifteen more, DD-95 to -109) were authorised in March 1917, and, drawing on a Naval Emergency Fund, which approved 'such additional Torpedo Boat Destroyers…as the President may direct', contracts for a total of 61 were placed by May, up to the *Tillman* (DD-135). After discussions about 'austere' designs, two variants of the existing flush-deckers were chosen, one from Bath Iron Works (the *Wickes* class) and the other from Bethlehem Steel for its Quincy and San Francisco yards (the *Clemson* class). There were many variations in machinery; Bath Iron Works used Parsons or Westinghouse turbines and Normand, Thornycroft or Yarrow boilers, whereas Bethlehem Steel used Curtis turbines and Yarrow boilers. As production expanded other yards were awarded contracts, but they built to the Bath Iron Works design. The record for building-time was held by the *Ward* (DD-139), commissioned at Mare Island Navy Yard only 70 days being laid down, but eight to ten months was more usual.

After the Armistice the US Navy was saddled with huge numbers of destroyers, and many were laid up. Several were earmarked for conversion to minelayers (DMs) and seaplane tenders (AVDs and AVPs), and when the Second World War broke out several more were converted to fast transports (APDs) and fast minesweepers (DMSs). Fifty were transferred to the Royal Navy in September 1940 under the famous 'destroyers-for-bases' deal, and took the names of towns common to the British Empire and the United States. Some were converted to banana-carriers between the wars, and the *Stewart* (DD-224) was captured by the Japanese in 1942, commissioned into the Imperial Japanese Navy and was returned to the US Navy in August 1945. One of the Royal Navy's 'Town' class, HMS *Campbeltown*, ex-USS *Buchanan* (DD-131) headed the St Nazaire Raid in March 1942, during which she was rammed into the large dry dock and blown up to deny the dock's use to the German battleship *Tirpitz*. The last 'flush-decker' in service appears to be the Russian *Druzni*, ex-HMS *Lincoln*, ex-USS *Yarnall* (DD-143), returned to the Royal Navy and scrapped in Britain in 1952.

The Royal Navy acquired fifty flush-deck destroyers as part of the 'Lease-Lend' programme in 1940. Though the design was some thirty years old, they scored a few successes against more modern U-boats.

Flush-deck destroyers

Caldwell class, 6 ships (DD-69 to-74), built 1916–17
Wickes class, 112 ships (DD-75 to-185), built 1917–19
Clemson class, 162 ships (DD-186 to-347), built 1918–22

Caldwell class

Displacement:	1020 tons (normal), (deep load)
Dimensions:	308-310ft (wl); 315ft 6in (oa) x 30ft 9in x 7ft 6in
Machinery:	DD-69-72: 3-shaft geared steam turbines, 21,000 shp; DD-72-74: 3-shaft geared steam turbines, 18,500 shp
Speed:	30kts; four Yarrow boilers
Armament:	4-4in 50 cal (4x1); 12-21in torpedo tubes (4 x 3) DD-73: 5-4in (1 x 2, 3 x 1) + 12–21in)
Range:	2500 nm @ 20kts
Fuel:	c. 205 tons oil
Complement:	5 officers and 95 enlisted personnel

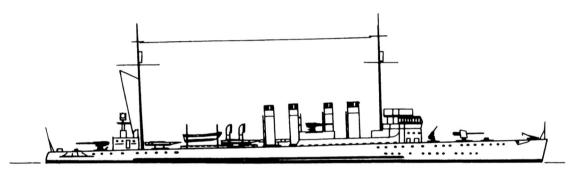

The destroyers of the Wickes class differed radically in range, a consequence of different engines, and the best of those built at the Bath Iron Works were designated 'Long Radius Boats'. (Line drawing 1/720 scale.)

Wickes class

Displacement:	1090 tons (normal), 1247 tons (deep load)
Dimensions:	310ft (wl), 314ft 3in (oa) x 30ft 11in x 8ft 6in
Machinery:	2-shaft geared steam turbines, 26,000 shp; 4 boilers
Speed:	35kts
Armament:	4-4in 50 cal (4 x 1); 12-21in torpedo tubes (4 x 3)
Range:	2500 nm @ 20kts
Fuel:	225 tons oil + 50 tons (emergency)
Complement:	6 officers and 108 enlisted personnel

Clemson class

Displacement:	1190 tons (normal), 1308 tons (deep load)
Dimensions:	310ft (wl), 314ft 4.5in (oa) x 30ft 11in x 9ft 3in
Machinery:	2-shaft geared steam turbines, 27,500 shp; 4 boilers
Speed:	35kts
Armament:	4-4in 50 cal (4 x 1), 1-3in AA or 2-1pdr (2 x 1) AA, 12-21in torpedo tubes (4 x 3)

Range: 2500 nm @ 20kts
Complement: 6 officers and 108 enlisted personnel

Conclusion

The 'flush-decker' programme, impressive though it was in sheer volume and building-times, only produced 39 destroyers in commission by the Armistice in November 1918. Many keels were not laid until 1920, two years after the Great War was over. The builders had warned in 1917 that such a large programme would yield meagre results, and very few after the *Greer* (DD-145) were completed in 1918.

Workmanship and performance varied from yard to yard, and this was reflected in endurance, so essential to convoy escort and anti-submarine tactics generally. The best performances were recorded by the Bath Iron Works and Cramp vessels, the worst by those built by Quincy. The difference resulted in the Bath-built destroyers being rated as 'long radius' ships to distinguish them from the 'short radius' ships.

In service in the Second World War the age of their hulls and machinery were inescapable weaknesses. The Royal Navy found that its 'Town' class were carrying too much topweight and were too flimsy for the North Atlantic. Despite the immense effort put in during the decade 1906 to 1916, the pace of improvements to machinery was sometimes counter-productive. In fact advanced technology was 'frozen' prematurely in an effort to boost endurance, a severe disadvantage when embarking on mass-production. According to the CO of HMS *Beverley*, ex-USS *Branch* (DD-197), a major weakness was the lack of 'preventer strips' on the bearings of the propeller shafts. These were hard metal longitudinal strips embedded in the soft metal of the bearing, intended as a 'get you home' solution if one or other of the main shaft bearings 'ran', i.e. melted. Unlike contemporary 'V & W' class destroyers, 'run' bearings could immobilise the ship.

For the US Navy the vast numbers of 'flush-deckers' in existence in the 1920s was an embarrassment. Congress and the Treasury, having funded the huge programme in 1917–20, was unsympathetic to requests for more modern destroyers. All that could be done was to use as many as possible on humdrum subsidiary duties until the political climate improved. For the Royal Navy in 1940–41 the 50 'Towns' were less of an embarrassment, but the long time spent in reserve had taken its toll on such important items as wiring, and they took some months to become operational.

With the benefit of hindsight, a more coherent destroyer-design policy up to 1916 might have produced better results, through evolution rather than revolution. The resources devoted to the *Wickes* and *Clemson* classes might have been better directed towards production of that 'austere' design that so many people had argued for. Unfortunately wartime emergencies have a bad habit of generating panic responses rather than correct ones.

K class Submarines

Royal Navy 1915–19

To anyone who has the slightest knowledge of naval operations in the First World War the story of the K class is a huge disaster, a flawed concept badly executed. To a layman the idea of a steam-driven submarine seems so ridiculous as to prove that the Royal Navy was hopelessly incompetent. To that extent the Ks may seem a soft target, but there is much more to their story.

Early in 1915 the Commander-in-Chief of the Grand Fleet, Admiral Sir John Jellicoe, asked the Admiralty to investigate the possibility of building a new category of 'fleet' submarines capable of accompanying the battlefleet to sea. This required a minimum speed of 21kts. Such submarines would in theory be able to ambush the German High Seas Fleet—at this time Jellicoe and his staff were haunted by the fear of being lured into a 'U-boat trap', and it seemed sensible to try something much more dangerous against the German fleet.

The first attempt, the J class, was only a qualified success. Despite everything the engineers could do, there was no diesel engine powerful enough for the task, and the best the new boats could achieve was 19kts. Vickers developed a 12-cylinder diesel, and one drove each of three shafts, but the desired speed could not be achieved. Then a rumour (false, like so many wartime guesses) reached the First Sea Lord, Admiral Fisher, to the effect that the Germans had produced a design for a 22kt U-boat, and Fisher insisted that any fleet design must go one better.

In April 1915 Vickers, the country's leading submarine-builder, submitted a design for a submarine driven by geared steam turbines and a diesel engine. This was unsolicited, but after some thought, a 1913 design was resurrected and the best features from each one were selected as the basis of a new design. The 18in torpedoes of the J class were chosen to save time, with two beam tubes as well as four in the bow, and a twin revolving pair were sited in the superstructure for use on the surface. This armament of ten torpedo tubes was exceptionally heavy for a 1915 submarine, and provision of beam tubes was to allow a beam shot without waiting for the submarine to be steered onto the correct bearing. Gyro angling of torpedoes was not to become available for many years. The gun armament was also heavy, two single 4in guns (one replaced by a 3in anti-aircraft gun, a sign of things to come). *K.17* was armed with a 5.5in gun.

To meet the Staff Requirement for a speed of 24kts the powerplant was a pair of steam turbines developing 10,000 hp, using steam generated by two Yarrow boilers. In the original design a diesel engine was intended to drive a centre shaft, but in the final version an 800bhp diesel generator drove a 700hp dynamo to supply current to the electric motors. This was the first example of diesel-electric drive in Royal Navy submarines. Two 700hp electric motors were coupled to each of two shafts. The intended diving procedure was to shut down the boil-

The K.3 before her bows were raised. This original configuration took a lot of water when steaming at speed in rough weather.

ers, lower the funnels into wells in the superstructure (closed by heavy caps), and then clutch in the electric motors and the 386-cell battery, as in a conventionally powered submarine.

The first orders (for K.3 and K.4) were placed in June 1915, an amazingly short period to develop the design, and a second order for twelve followed in August. Seven more were ordered in 1916, but four orders were cancelled. When the first boats were delivered, problems took two forms. First, they were very complex and much larger than any existing design. The ease with which they could operate in all weathers with surface ships steaming at high speed was badly overestimated; the view from the conning tower was limited because it was so close to the water, and the submarine itself was equally difficult to spot from surface ships. It was said by a K boat officer that 'they handled like destroyers but had the bridge facilities of a picket boat'. A K was 338ft long, and if mistaken for a U-boat by ships of its own fleet was simply not manoeuvrable enough to take avoiding action.

Getting the two diminutive funnels folded down into their watertight wells took only half a minute, but there were also a number of quadruple mushroom-topped ventilators to be secured, and a small obstruction such as a strop of wire rope was sufficient to prevent them from closing properly. With nobody on deck at the time when preparing to dive, everything depended on indicators inside the pressure hull. A problem rarely understood by modern commentators is that early submarines were virtually no more than submersible torpedo boats, and were intended to operate at little more than periscope depth. The 'crush-depth' was usually no more than 150 per cent of the length of the submarine, and a K diving at speed was very difficult to control. Her speed was closer to the surface speed than the 9.5kts on electric motors, and she could all too easily go beyond her crush-depth before she began to respond to the diving planes.

The early Ks had a nearly flush bow, and it is clear from photographs that they took a lot of water when steaming at speed in rough weather. To deal with this the designers had provided an enclosed wheelhouse, but it was too low to provide good visibility. To cure this problem they were given a prominent 'swan bow' containing a quick-blowing tank. It improved seakeeping, and the quick-blowing tank helped to bring the bows up after a fast dive. To provide bridge personnel with a field of view over the swan bow the conning tower was raised, giving the Ks a very distinctive profile. Another pointer to problems with seakeeping is the number of variations in the positions of deck guns; they were moved about on deck and frequently shifted to the superstructure, and K.12 had a prominent raised cutwater for her 4in gun, just forward of the heightened conning tower.

As they were delivered from their builders the boats were allocated to the 12th and 13th Submarine Flotillas, which formed part of the Grand Fleet organisation. Although the K class were involved in a number of accidents, talk of a 'jinx' on the class should not be exaggerated. In all 17 were built, of which 5 were lost in 1917–21:

K.13 sank during trials in the Gareloch on 29 January 1917 but she was raised two months later and returned to service as *K.22*
K.1 was sunk by the guns of the light cruiser HMS *Blonde* following a collision with her sister *K.4* in poor visibility off the coast of Denmark on 17 November 1917
K.4 was sunk in collision with *K.8* in Firth of Forth on 31 January 1918
K.17 was sunk in collision with the light cruiser HMS *Fearless* in the Firth of Forth on 31 January 1918
K.5 was lost in a diving accident in the Bay of Biscay on 20 January 1921

The biggest blot on their reputation was undoubtedly the so-called 'Battle of May Island' in the Firth of Forth at the end of January 1918, in two flotillas of four and five respectively. While steaming at high speed without lights, *K.14*'s helm jammed and she was rammed by *K.22*, throwing the line into confusion. A squadron of batlecruisers was coming up astern and steamed straight through the stricken flotilla, HMS *Inflexible* striking *K.22* a glancing blow without realising what had gone wrong, The flotilla's leader, the light cruiser HMS *Fearless*, turned back to find what had happened, and rammed the crippled *K.14*, cutting her in half. In the confusion that followed, *K.4* was rammed and sunk. *K.17* was sunk, and *K.8* was badly damaged by *K.6* and *K.17*. The whole affair was traumatic for the crews of the class, with over 100 of the flotilla-mates dead, and the inevitable news blackout simply enhanced the credibility of rumours. A nod in the direction of superstition resulted in the renumbering of *K.13* to *K.22* after her sinking in the Gareloch in January 1917 during trials, but believers in the malign fate suggested by that 'unlucky' number will be comforted by the fact *K.22* started the train of events in the 'Battle of May Island' a year later by ramming *K.14*.

The Court of Enquiry was critical of some of the K boats' officers, but with the benefit of hindsight it is hard to see how a sharper lookout or quicker reactions could have avoided disaster. For reasons of operational secrecy no navigation lights or radio signals were permitted.

The speed with which the design was rushed into service meant that a number of detailed design faults crept in, and early in 1918 six of an improved design were ordered, *K.23-28*. Five were cancelled after the Armistice, but *K.26* was completed in 1923. As her hull was larger with the same powerplant, speed was slightly reduced, but she was armed with the much more effective 21in torpedo. The guns were mounted on the superstructure, and were protected by revolving shields. She proved a success and served for eight years, undertaking a round-the-world trip among other achievements.

K CLASS SUBMARINES

K.1	launched 14 November 1916	Portsmouth Dockyard
K.2	launched 14 October 1916	Portsmouth Dockyard
K.3	launched 20 May 1916	Vickers
K.4	launched 15 July 1916	Vickers
K.5	launched 16 December 1916	Portsmouth Dockyard
K.6, K.7	launched 31 May 1916	Devonport Dockyard
K.8	launched 10 October 1916	Vickers

K.9	launched 8 November 1916	Vickers
K.10	launched 27 December 1916	Vickers
K.11	launched 16 August 1916	Armstrong
K.12	launched 23 February 1917	Armstrong
K.13/K.22	launched 11 November 1916	Fairfield
K.14	launched 8 February 1917	Fairfield
K.15	launched 30 October 1917	Scotts
K.16	launched 5 November 1917	Beardmore
K.17	launched 10 April 1917	Vickers
K.18 and *K.19*	ordered but cancelled 1916 before being laid down	
K.26	launched 26 August 1919	Vickers (completed by Chatham Dockyard)

K.1-22

Displacement:	1880/2650 tons (surfaced/submerged)
Dimensions:	338ft (oa) x 26ft 6in x 16ft
Propulsion:	2-shaft geared steam turbines, 10,000shp; 4 electric motors, 1400hp; diesel 800bhp
Speed:	24/9.5kts (surfaced/submerged)
Armament:	10-18in TT (4 bow, 4 beam, 2 revolving above water); various combinations of 5.5in, 4in and 3in AA guns
Range:	3000nm @ 13.5kts
Complement:	59

K.26-28

Displacement:	2140/2770 tons (surfaced/submerged)
Dimensions:	351ft 6in (oa) x 28ft x 16ft 6in
Propulsion:	As K.1-22
Speed:	23.5/9kts (surfaced/submerged)
Armament:	8-21in TT (4 bow, 4 beam); 3-4in guns
Range:	As K.1-22
Complement:	60

CONCLUSION

The K class are very unusual, in that virtually all commentators have ridiculed the concept of a steam-driven submarine, while totally ignoring the operational concept and the technical ingenuity of the design. Steam propulsion was not a new idea, having been used in a series of French submarines before 1914. Diesel technology could not deliver the power needed for 24kts surface speed, and it is therefore hardly surprising to find that a proposed 1917 design for a German fleet U-boat (Project 50, the UD.1 class) also stipulated steam propulsion. There is nothing inherently wrong with steam propulsion for submarines—every nuclear boat since the USS *Nautilus* has been driven by steam turbines. In the K design special attention was paid to ensuring a rapid and safe shutdown of the steam plant before diving, to avoid the lengthy procedure necessary in the French 'steamers'. The drawback, however, was that the high-speed plant required large quantities of air to provide sufficient draught for the boilers; that meant large ventilators, each set of which had to be closed as tightly as possible when diving. Any failure of procedure, whether through human error or unavoidable cause, was potentially fatal.

K.26 *was by far the most successful member of the K class. She was completed after the war, having been ordered in early 1918.*

The list of losses does not include a number of non-fatal collisions, all of which have been an important element of the mythology surrounding the Ks. But any research into ship accidents in the First World War shows that all classes of warship, from battleships down to minesweepers, were involved in collisions. The Royal Navy played a particularly aggressive role to deny sea control to the Imperial German Navy, and this meant offensive operations in enemy waters in all weathers. In those pre-radar days ships steaming fast without lights took risks constantly. In poor visibility collisions were bound to happen, so the loss of **K.1** and even the two off May Island cannot be blamed on the design. The loss of *K.13* on trials and the unexplained accident which caused the loss of *K.5* were matched by similar losses to all submarines before 1917 and for many years after.

The tactical value of fleet submarines has been generally criticised, particularly by submariners. The true role of the submarine, they argue, is that of the lone predator, operating well away from friendly warships. But there was a rationale for the K class. They were intended to accompany the battle fleet until an engagement was imminent, and then be detached to form a 'submarine trap' to ambush enemy units and to pick off damaged stragglers. For that they needed a margin of speed over the battlefleet, and, given the relatively inefficient state of ship-to-ship radio communications, they had to remain close to the battlefleet to maintain tactical cohesion. One of the 'might-have-beens' of naval history is what might have happened if a flotilla of four or five 'K' boats had been available at the Battle of Jutland in May 1916. One can only speculate what might have happened to Admiral Scheer's battered High Seas Fleet if it had encountered five Ks off the Horns Reef at the end of its run for home.

The K design was a giant step in the evolution of the submarine, but its huge increase in size made for very delicate handling when submerged. The crush-depth of the Ks was about 500ft. The science of metallurgy was well developed, but the behaviour of steel under stress was not properly understood, and it was unlikely that a K diving to 400ft (all too possible with such a long hull) would behave exactly as predicted.

All in all, the K class submarines were a bold attempt to give the submarine a new tactical role, integrating it with surface operations—a dream which is only being turned into reality with the latest nuclear attack submarines (SSNs) and advances in communications. The designers and the Naval Staff cannot be blamed for making the effort eight decades earlier.

HM Ships *Courageous*, *Glorious* and *Furious*, Light battlecruisers

Royal Navy 1915–48

Very shortly after war broke out in August 1914 the British Cabinet ordered all work on capital ships to be stopped. It was, after all, going to be a short war, and the Royal Navy's current programme would add to the already big preponderance over Germany's High Seas Fleet.

The Cabinet authorised the Navy's political head, First Lord of the Admiralty Winston Churchill, to recall the retired Admiral Fisher as a replacement for the 'too German' First Sea Lord Prince Louis of Battenberg at the end of October 1914. Little did the politicians realise that they had unleashed a tiger who would try to run the Royal Navy as *he* wanted. Fisher set about creating a vast shipbuilding programme. Despite his advanced years he displayed all the ferocious energy of his previous tenure of the post. In addition to a huge force of submarines, destroyers, shore bombardment ships and light cruisers, Fisher wanted to build new and more powerful battlecruisers. Despite the flaws in the original battlecruisers, his ideas remained unchanged; speed was the key to tactical and strategic success, and armour took second place. In wartime he could advance these ideas without too much political interference; resources rather than cost drove the procurement process.

Using the victory in the Battle of the Falklands, Fisher urged the senior fleet commanders and the politicians to allow him to restart battlecruiser-construction. Churchill, an ardent admirer of Fisher, caved in to the massive lobbying effort, and at the end of December 1914 obtained Cabinet approval to build two new battlecruisers (but not the three that Fisher had demanded). Using as much of the material as possible from the steelwork already done for two suspended battleships, he produced two 32,000-ton ships, *Renown* and *Repulse*. They were armed with three twin 15in gun mountings each, all that was available in the short term, but were protected only by 6in belt armour, no more than the original battlecruisers designed a decade earlier. He was oblivious to the vastly greater risk of serious damage from the new generation of bigger guns such as the German 38cm and his own Navy's 15in.

Fisher was not satisfied with this victory of prejudice over practicalities, and looked for other ways to evade the Cabinet prohibition on more battlecruisers. The result was a new concept of 'large light cruisers', ships with very light protection, high speed and a small number of heavy guns. It was widely believed at the time that these 18,000-ton ships were intended to be part of Fisher's vaunted Baltic Project, a large-scale seaborne invasion of the Baltic to deny the High

HM Ships Glorious, *her sister* Courageous *and her half-sister* Furious *were all products of Fisher's second coming to the Admiralty.*

Seas Fleet its 'sanctuary' and to support a Russian landing. He had talked about such a plan many years earlier but nothing has been found in the way of planning papers, and in the opinion of one biographer, it was no more than a topic for dinner-party conversations!

Assuming that the Baltic Project was never more than a pipedream, what was the purpose of these 'large light cruisers'? Although he told Churchill and the Commander-in-Chief of the Grand Fleet, Sir John Jellicoe, that they were intended to support the Baltic Project, he was marginally more candid with the DNC, Sir Eustace Tennyson d'Eyncourt. Fisher wrote in March 1915 that he saw the ships as the fulfillment of his ideals, 'all three requisites of gunpower, speed and draught of water so well balanced'. He also said that he envisaged them hunting down enemy cruisers on the high seas. Ninety years later, the impression is one of exuberance about the technical feat of building such ships rather than much serious thought about their use.

The 'legend', outlining the allocation of weights for the first two, was submitted for approval at the end of January 1915. They were to have the following characteristics:

1 Sufficient displacement to ensure high speed in rough weather.
2 Draught restricted to allow operations at the entrance to the Baltic; it was to be about 5ft 6in below the average for capital ships, and all other factors were subordinate to this requirement.
3 A powerful armament.
4 A speed of at least 32kts.
5 Protection to be of light cruiser standard.
6 Underwater 'bulges' for a considerable length, and machinery and boilers protected by being as far inboard as possible behind triple bulkheads.

The design was, in essentials, a reduced version of the *Repulse* and *Renown*, but with side armour

THE WORLD'S WORST WARSHIPS

The Admiralty chose to convert HMS Courageous *into an aircraft carrier after the signing of the Washington Naval Treaty in 1922.*

on the scale of existing light cruisers, i.e. 3in. On the propulsion side, however, they benefited from a more efficient powerplant, with small-tube boilers and geared turbines. Fisher took a great interest in these 'perfect' ships, and early in March told the DNC to thicken the 0.75in torpedo (protection) bulkheads. They were to be named *Courageous* and *Glorious*; both names were used for the first time, although French and Spanish variants were used in the days of sail.

The DNC pointed out that the change would add to the displacement, but Fisher overrode his objections, and the torpedo bulkhead was increased to 1.5in. This added 500 tons and reduced speed by 0.25kts. Fisher's departure in his celebrated huff during the Dardanelles Campaign brought an end to his control over the construction, probably a good thing. The only later major addition was thicker deck plating around the bases of the turrets after the Battle of Jutland, but cumulative small changes added another 400 tons to the displacement. Although there is no record of any other additions both ships were 1700 tons heavier than the original legend figure. Both took about 18 months to build, *Courageous* starting trials in October 1916, followed by *Glorious* two months later.

The first set of trials of *Courageous* showed the risks of driving a light hull in bad weather. While working up to maximum speed against a rough head sea she suffered structural damage: buckling of the deck and side plating of the forecastle, and leakage in oil bunkers and boiler feedwater tanks. An investigation was unable to decide whether the damage was caused by her light construction or by being driven too hard in rough weather. The cure was another 130 tons of stiffening, which also approved for her sister, although in her case it was not done until 1918.

The armament comprised two twin 15in guns, the largest then available, and a secondary battery of six triple 4in guns and a pair of 3in anti-aircraft guns. The two ships were involved in a scrappy light-cruiser action in the Heligoland Bight in 1917, the only occasion on which they fired their guns in anger. HMS *Courageous* spent some time as a minelayer in 1918, but apart from that, they so no action again. By a strange coincidence the guns themselves did fire again, although not in anger. The four turrets and their guns were 'mothballed' and were selected in 1940 to arm a projected battleship, which was completed as HMS *Vanguard* in 1946.

After the Washington Naval Disarmament Treaty came into force in 1922 these strange hybrids had to be discarded because their 15in armament put them in the category of capital ships. Their guns were removed and in leisurely fashion both were converted to aircraft carriers. This career-move proved unlucky, as they were early losses in the Second World War. HMS *Courageous* was torpedoed by a U-boat in the Western Approaches three weeks after the outbreak of war, and HMS *Glorious* was sunk by the gunfire of the German battlecruisers *Gneisenau* and *Scharnhorst* at the end of the Norwegian campaign in June 1940.

The *Furious* was built to a modified design, with a different form of torpedo protection and two single 18in 40-cal guns with an elevation of 30 degrees. This was a giant step forward in gunnery terms, and demonstrates the capacity of the naval sector of industry. The 15in 42-cal gun had proved so successful that the designers merely expanded the design, to produce the most powerful gun in the world at the time firing a 3320lb shell. As an insurance against failure the barbette diameter was the same size as that for the 15in turret, allowing a reversion to 15in twins if the 18in gun was a failure. The 18in gun proved a great success, however, with long range (30,000 yards) and accuracy.

HMS Furious's *future as an aircraft carrier was already planned even before she was completed. In 1918, she participated in the successful air raid on the Zeppelin base at Tondern.*

Ordered in 1915 from Armstrong's, the prototype was tested in 1916, and three barrels were ready in 1917, a very short development time. But, without Fisher's driving force, the ship was earmarked for conversion to an aircraft carrier before she was completed. The forward gun was delivered to the shipyard but never fitted, and a hangar and sloped flying-off deck were built in its place. Naval aviation was advancing at a hectic pace, and in 1918 she was given a further conversion to allow aircraft to land abaft the funnel on a long landing deck. She made history in August 1917, when Squadron Commander Dunning RN successfully landed on the flying-off deck. Yet another, much more comprehensive reconstruction followed 1922–1925, turning her into a fleet carrier. As such she had a successful career in the Second World War. Her contribution in the previous conflict had been limited to the successful Tondern raid in 1918, but it marked the dawn of a new era.

As a footnote, the three 18in guns were allocated to rearm three of the Dover Patrol monitors for seaborne bombardment of the Belgian coast, although only two ships were converted by November 1918. For years it was widely believed that the guns were subsequently mounted in the coast defences of Singapore, but it has been proved that two were sold for scrap in the early 1930s. The surviving barrel was used for a variety of gunnery trials at Shoeburyness until after the Second World War.

HM Ships *Courageous* and *Glorious*

Courageous laid down 28 March 1915, launched 5 February 1916, completed 28 October 1916, built by Armstrong's, Newcastle on Tyne

Glorious laid down 1 May 1915, launched 20 April 1916, completed 14 October 1916, built by Harland & Wolff, Belfast

Displacement:	*Courageous*: 17,400 tons (designed), 22,560 tons (deep load)
	Glorious: 22,360 tons (deep load)
Dimensions:	735 ft (pp), 786ft 9in x 81ft (moulded) x 25ft 10in (max)
Machinery:	4-shaft 90,000 shp Parsons geared steam turbines; 18 Yarrow small-tube boilers
Speed:	32kts (designed), 31.5kts (actual)
Armament:	4-15in Mk I 42 cal. guns (2 x 2), 18-4in Mk IX (6 x 3), 2-3in AA (2 x 1), 2-3pdr (37mm) (2 x 1), 2-21in torpedo tubes (submerged)
	Courageous 222 Elia mines or 202 HII type (as minelayer 1918)
Armour:	2in HT steel with 1in skin, 1–3in deck, 3–6in barbettes, 3–10in CT
Fuel/range:	750 tons (normal), 3250 tons (maximum)/6000 nm @ 20kts
Complement:	768–787 officers and ratings

HMS *Furious*

Laid down 8 June 1915, launched 18 August 1916, completed 26 June 1917, built by Armstrong's, Newcastle on Tyne

Displacement;	19,513 tons (designed), 22,890 tons (deep load)
Dimensions:	735ft (pp), 786ft 9in (oa) x 88ft x 24ft 7in
Machinery:	4-shaft 90,000 shp Brown-Curtis geared steam turbines; 18 Yarrow small-tube boilers
Speed:	31.5kts
Armament:	2-18in Mk I 40 cal guns (2 x 1, designed), 11-5.5in 50 cal QF Mk I (11 x 1), 2-3in AA (2 x 1), 2-3pdr (37mm) (2 x 1), 2-21in torpedo tubes (submerged)
Armour:	As *Courageous*
Fuel/range:	750 tons
Complement:	726 (as designed

Conclusion

There is no doubt that these three hybrid ships were never going to be successful in their intended role. The fire-control problems of achieving accurate long-range fire from only two turrets were twice as difficult as those encountered in contemporary battleships, which were scoring only 2 per cent hits. This was pointed out by the Director of Naval Ordnance at the design stage. And if that was true of *Courageous* and *Glorious*, how much worse would it be with the two single 18in planned for *Furious*?

The indifference of Fisher to this basic criticism of his creations reflects the sad truth that he was out of his element and too old for the job. His views on naval warfare were as naive as those of mid-Victorian pundits, in assuming that a hit from a single large-calibre shell would have a cataclysmic effect. He boasted that the guns of HMS *Furious* 'with their enormous shells were built to make it impossible for the Germans to prevent the Russian Millions from landing on the Pomeranian Coast'. He went on to predict that a single hit from a 20in gun (intended to follow the 18in for another battlecruiser design) would produce a crater 'like that of Vesuvius or Mount Etna', sending the German Army 'fleeing for its life from Pomerania to Berlin'.

Fisher seemed to be unaware that shells landing out of sight were, by definition, not accurate. Nor was there the remotest possibility that a small number of 18in shells would produce volcano-sized craters or scatter the German Army. In a stern chase against enemy light cruisers the end-on fire of a twin 15in turret would be inaccurate, while the accuracy of a single 18in gun in the same circumstances is hardly worth considering. In the case of the *Furious*, the tremendous muzzle-blast of her single 18in gun 'shook the ship up considerably', according to an observer.

Once the Baltic Project was buried very real doubts were voiced about the best way to employ the ships. In June 1916 the DNC's office was asked about the cost and time penalties and the complexity of converting *Courageous* and *Glorious* to seaplane carriers. As fast escorts to light cruisers and destroyers they showed some promise, but were an expensive solution to that problem. In 1917–18 they were fitted with a dozen additional twin 21in torpedo tubes, one pair port and starboard of the mainmast, at upper deck level, and two pairs port and starboard on the quarterdeck, flanking the 15in turret. In 1917 *Courageous* was fitted as a minelayer, with mine rails laid on the quarterdeck, and chutes fitted at the stern, but she never laid a mine, and the equipment was removed in 1918.

With the wisdom of hindsight, it can be argued that a better use of the first pair would have been as escorts for fast aircraft carriers. But that would not have got around the prohibitions of the Washington Treaty, and any attempt to rearm them with cruiser-sized guns would have proved too costly. As carriers they were at least moderately successful, although they suffered from relatively small aircraft capacity.

Spurious, *Curious* and *Outrageous*, as sailors dubbed the ships, demonstrate the folly of designing warships without much serious thought about their roles. In that the trio of white elephants merely followed the tradition established by Fisher for his original battlecruisers, and they contributed to the general disillusionment with his alleged 'genius' as a designer of warships. Of course, Fisher was not a designer, merely an explosively persuasive driving force that dragged lesser mortals along. The saga gives an insight into the minds of contemporary senior officers and Admiralty civilians, none of whom seemed to summon up the courage to tell the old admiral that he was wrong. On the other hand, Fisher's vindictiveness towards anyone who tried to contradict him was already well known.

Fast Battleship HMS *Hood*

Royal Navy 1916–1941

Many may question the description of HMS *Hood* as a fast battleship rather than a battlecruiser, but that is how she was conceived. In a note in late 1915 the Controller suggested the construction of an 'experimental battleship' armoured, armed and engined on similar scale to the then new Queen Elizabeth class. She was, however, to have, if possible, a 50 per cent reduction in draught and incorporating the latest ideas on underwater protection. Shallow draught was intended to reduce the hydrostatic pressure on bulkheads and to provide a bigger reserve of buoyancy. Greater freeboard would also prevent the secondary armament from being washed out in rough weather.

On 29 November the DNC, Sir Eustace Tennyson d'Eyncourt, produced a rough draught for a ship 760ft long, with a beam of 104ft and a draught of only 23ft 6in (a 22 per cent reduction over the Queen Elizabeth design). Although the 10in belt was thinner than the Queen Elizabeth's, it was sloped, and reckoned to offer equivalent resistance. These proposals (and inferior alternatives) were forwarded to Sir John Jellicoe, Commander-in-Chief of the Grand Fleet, for his comments.

Jellicoe replied that there was little need for more battleships, as the Grand Fleet already had a handsome margin over the German High Seas Fleet, but he needed faster ships, especially as intelligence suggested that the Germans were building three battlecruisers capable of 30kts and armed with 38cm guns (the Ersatz Yorck class, scrapped incomplete in 1918). Several solutions were prepared to meet the new requirements, all large ships, to be armed with the existing 15in gun or the new 18in, which had started development. In the event a design with eight 15in guns in four twin turrets was accepted, with the 10in armour reduced to 8in in the discarded designs.

The final design was accepted in March 1916:

Displacement:	36,300 tons
Dimensions:	860ft x 104ft x 25ft 6in
Armour:	8in belt, 5in upper belt, 4–5in forward, 3–4in aft
Machinery:	144,000 shaft horsepower
Speed:	32kts
Oil fuel:	1200 tons at load, 4000 tons

Three ships were ordered on 17 April 1916, with a fourth in July. The *Hood* herself was laid down on 31 May 1916, a bad omen, for it was the date of the Battle of Jutland, during which three lightly armoured battlecruisers were sunk with heavy loss of life. Work on the new battlecruisers was suspended immediately to give time for studies into protection, anti-flash precautions and projectile design.

As a result a series of piecemeal improvements was begun, notably deepening the 8in belt, slight increases in deck armour and heavier armour on the 15in turrets. These changes would have pushed displacement up to 37,500 tons and would have reduced speed by 0.25kts. On the same day that these were put in hand, 25 June 1916 (just over three weeks after Jutland) the DNC offered a much more radical proposal. This harked back to the original concept of a fast battleship, with 12in side armour, 6in upper side armour and 15in on the barbettes supporting the turrets, all for a loss of only 1 knot. The Admiralty Board accepted this development and re-design started on 1 September, but the design was not finally accepted until August 1917, after a series of minor improvements had been made; displacement went up by 600 tons. The sloped 12in side armour was the equivalent of 14–15in, making her the best protected capital ship of her day, but her deck armour was not up to the same standard. Nor did the minor postwar changes make effective use of the tonnage available. Had HMS *Hood* been designed after the war, when the lessons had been learned, she would have been armoured on the 'all or nothing' principle pioneered in the USS *Nevada* in 1911 and adopted for the 1920-21 series of designs which culminated in the protection system for HMS *Nelson* and *Rodney* (laid down in 1922).

A breakdown of armour weights as a percentage of displacement shows that the ship was indeed a fast battleship rather than a classic battlecruiser:

Invincible	20%
Hood	33%
Nelson	29%
King George V (1936)	36%
Hindenburg	34%
Lexington	28.5% (as battlecruiser, 1919)

(Comparisons are approximate. No two navies counted weights in the same way.)

The Hood incorporated a number of changes that saved weight compared with pre-1914 battleships, but was still built before all the lessons of the First World War had been fully understood. (Line drawing 1/2400 scale.)

HMS *Hood*

Laid down: 31 May 1916, launched: 22 Aug 1918, completed: 15 May 1920
Built: John Brown & Co, Clydebank

Displacement:	42,670 tons (load)
Dimensions:	810ft 5in (pp) x 104ft 2in x 29ft 3in (mean)
Machinery:	4-shaft geared steam turbines. 144,000 shaft horsepower, 24 boilers
Speed:	31kts
Armament:	8-15in 42 cal (4 x 2), 12-5.5in 50 cal guns (12 x 1), 4-4in Mk V single AA guns (4 x 1), 6-21in torpedo tubes (submerged, 4 above water)
Complement:	1397

THE WORLD'S WORST WARSHIPS

When the *Hood* was launched at the Clydebank yard of John Brown in August 1918, no-one sensed any ill omens, but with hindsight we cannot ignore an unfortunate coincidence. The new ship's sponsor was Lady Hood, whose husband, Rear Admiral Horace Hood, had died at Jutland when his flagship, the battlecruiser *Invincible*, blew up with only three survivors, the same fate as the *Hood* 25 years later.

But at her launch HMS *Hood* was the largest warship ever built, and her elegant lines inspired worldwide admiration. She was ideal for the purpose of reminding the world that the security of the British Empire rested on the world's largest navy. The 'Mighty Hood' began a series of official and 'flag-showing' cruises and visits to Scandinavia and South America, to the Mediterranean and the Pacific. In 1923–24 she led a round-the-world cruise, in company with the smaller battlecruiser HMS *Repulse* and five cruisers. The squadron visited South Africa, Zanzibar, Ceylon (now Sri Lanka), Singapore, Australia, New Zealand, the Pacific islands, Hawaii and San Francisco, then passed through the Panama Canal and called at Jamaica, Canada and Newfoundland before heading for home.

HMS Hood *photographed while serving as part of the non-intervention patrol during the Spanish Civil War.*

She was ageing, as all ships do, and if unmodernised would have been due for replacement in 1941. The alternative, a major modernisation, was warmly supported by the DNC, but the 1930s saw an increasing number of political crises in Europe, as the dictators became more confident in their ambitions. With an urgent programme of modernising the First World War battleships of the Queen Elizabeth class and the poorly protected battle cruisers *Renown* and *Repulse*, HMS *Hood* and the modern battleships *Nelson* and *Rodney* were the Royal Navy's main strength. Inevitably she was low on the list of priorities for upgrade, despite warnings from the DNC. In 1938 he warned that HMS *Hood* was in poor mechanical condition and that her thin deck armour made her 'unfit for front line service'. He went on to propose a number of new armour-schemes but concluded sadly that Treasury limits on expenditure ruled out any chance of reconstruction. In an uncannily accurate prophecy he warned, 'We may have eternal cause for regret.' Changes were limited to the addition of modern twin 4in (102mm) anti-aircraft guns at the outbreak of war.

The accretion of weight since the ship was launched in 1918, and the numerous wartime additions had reduced freeboard to such an extent that in rough weather she resembled a 'half tide rock'. This had no bearing on her loss, but combined with the parlous state of her main machinery, it reduced her maximum speed to 28kts. The only cure was to reduce weight, something that was virtually impossible in wartime.

From the outbreak of war the *Hood* took part in the monotonous round of patrols and 'sweeps'. To reduce weight her secondary 5.5in guns were removed in 1940, but the anti-aircraft guns were supplemented by a number of so-called UP (unrotated projectile) multiple rocket launchers on the shelter deck. A more useful addition was the installation of the new Type 284 gunnery control radar to improve the performance of the main armament. At last on 3 July 1940 she fired her 15in guns for the first time in war, when she opened fire on the French Fleet at Mers-el-Kebir, that 'melancholy action' to prevent the French ships from falling into Italian hands. Thereafter she returned to the Home Fleet as flagship of the Battlecruiser Squadron.

On 23 May, flying the flag of Vice Admiral Holland, HMS *Hood* and the newly completed battleship *Prince of Wales* were patrolling southwest of Iceland when, at 19:39, a signal from the heavy cruiser *Suffolk* was received, reporting a sighting of the German battleship *Bismarck* and the heavy cruiser *Prinz Eugen* in the Denmark Strait. Fifteen minutes later Holland ordered his force to change course to intercept the German ships and to increase speed to 27kts. At 04:00 the next morning they were steering a course of 240 degrees and steamed moved at 28kts, and at 05:10 both crews were ordered to Action Stations.

At 05:35 HMS Prince of Wales sighted the enemy at 17 miles (30,000 metres) and course was altered to starboard in order to close the range. Holland has been criticised for this end-on approach, which meant that the two British capital ships could only bring their forward guns to bear. But he wanted to close the range, and was aware that the Hood would become less vulnerable to damage as the range shortened and the trajectory of the German shells flattened. At 05:50 Hood and her consort made a further alteration of 20 degrees and *Hood* opened fire with A and B turrets, the *Prince of Wales* opening fire five minutes later. The German ships had very similar silhouettes, and *Hood* fired mistakenly at the *Prinz Eugen*, but *Prince of Wales* fired correctly at the *Bismarck*.

As the *Bismarck* fired her first salvo the two British ships altered course 20 degrees to port, probably to compensate for the German ship crossing from starboard to port. The salvo landed ahead of the flagship, and the second landed astern, but the third salvo straddled her and a hit started a fire on the port side of the shelter deck close to the mainmast (it could also have been an 8in hit from the *Prinz Eugen*). This hit set UP and 4in ammunition lockers on fire, and the flames spread forward and aft, but died down after a few minutes. *Bismarck*'s fourth salvo fell short, but the fifth straddled the *Hood*, and one or two hits were scored. Instantly a sheet of flame shot up in the vicinity of the mainmast and the *Hood* was almost totally enveloped in a huge cloud of smoke. When it cleared she had been blown in two; the after part sank quickly but the forward part reared up at an angle of 40 degrees, slid back into the sea and disappeared in three minutes. Nothing could be done to rescue any survivors at the time, and after the action the only living survivors found were a midshipman and two ratings, out of a complement of 1419.

According to a German survivor of the action, the massive explosion seemed to happen in silence, and the two German ships temporarily ceased fire as their bridge staffs tried to comprehend the catastrophe. The pride of the Royal Navy had vanished in the twinkling of an eye, in a manner all too reminiscent of the three battlecruisers at Jutland 25 years before.

Conclusion

The Admiralty was understandably anxious to find the cause of the catastrophic loss of HMS *Hood*. The public, despite wartime censorship, was told immediately, but the speculation was ill-informed, to say the least. Nor did the Board of Enquiry do much better, noting that the after 15in magazines had detonated, and concluding that a shell or shells had penetrated them or had set off 4in ammunition stowed nearby.

A second Board of Enquiry was appointed, this time with technical assistance from the DNC's staff. The DNC, now Sir Stanley Goodall, underlined the basic problem: the age of the *Hood* and the 1916 scale of deck protection. One of his staff summed it up succinctly: sending the *Hood* into action against the *Bismarck* was the equivalent of sending a late Victorian pre-dreadnought battleship into action at Jutland—the time-interval was similar and the outcome in both cases virtually inevitable.

There was a reason for the thin deck armour. Before Jutland the main fear was of hits from high-explosive HE shell at relatively shorter ranges, so attention was paid to keeping such shells out. Very long range plunging shells were not seen as a threat, not least because the propellant specialists had assured the DNC's designers that, if a magazine were hit from above, the stored modern cordite charges would burn without generating flash. Assuming that to be true (and it is difficult to see how the constructors could have challenged such assumptions), a hit penetrating a turret or barbette was believed to pose a very low threat. Jutland showed how false these premises were, but 1919 plans to increase deck armour over the magazines (5in forward, 6in aft) were never implemented. Even more ambitious plans to re-engine the ship and use the weight saved to re-armour her in 1938 fell foul of Treasury cost-limits.

The theory of a shell penetrating the after magazines ignored the puzzling fact that the large flash seen was far forward of the magazines. Sir Stanley Goodall, who had worked on the design as a Constructor, was convinced that the cause of the loss lay elsewhere. When the ship was under construction there was a great vogue for installing above-water torpedo tubes in capital ships, and in addition to the pair of underwater single tubes forward, four fixed 21in tubes were installed below the forecastle deck abreast of the mainmast (two firing to port and two to starboard). Goodall was convinced that a hit detonating the warheads on the port side (plus those of the reloads stored nearby), combined with the effects of the previous hit and the fire, could easily cause the ship to break in two.

Goodall's theory does not explain the detonation of X and Y magazines, but hits by two shells cannot be ruled out. The second shell may have penetrated either of the magazines, setting both off, or it could have detonated the anti-aircraft ammunition stored outside the armoured barbettes below the turrets.

We may never know all the answers, but at the time of writing (July 2001) an expedition has found the wreck of the *Hood*. Experience with other catastrophic losses suggests that the hull is so badly damaged that detailed analysis will prove impossible, but even establishing the position of the wreck has established a datum point for calculating the positions and movements of other ships involved in the action. A remotely operated vehicle (ROV) has located and examined the wreck, which lies at a depth of about 5,000 feet (1,500 metres) at position 63° 20' N, 31° 50' W. Previously the UK Ministry of Defence listed the wreck as a war grave and forbidden any attempts to examine it, but the unique historical importance of the loss of the Hood has resulted in the ban being lifted.

OMAHA CLASS SCOUT CRUISERS

US NAVY 1916–1949

After the Spanish-American War in 1898 the US Navy's interest in cruisers revived. Although the new armoured cruisers were virtually 2nd class battleships, another type emerged from 1903. This was the scout cruiser, intended to work with the battlefleet, relying largely on speed and seakeeping for protection, and providing intelligence on the whereabouts of enemy ships.

Despite the interest, only 13 scouts were built, three Chester class, authorized in 1904 and the ten Omaha class of the 1916 programme, which form the subject of this study. The reason was that Congress preferred to fund battleships, and as a result destroyers were forced to undertake scouting duties. The General Board kept asking for approval to build scouts (four in 1907, four in 1909, four in 1910 and four in 1911) but its submissions were ignored. In 1912 a decision was made to build two battlecruisers in Fiscal Year 1913. This was, however, abandoned in Fiscal Years 1914 and 1915, but reinstated in 1914 for the Fiscal Year 1916 programme.

By this time, reports of early operations in European waters emphasized the value of light cruisers. The General Board called for six scout cruisers and the Navy Department requested three for the Fiscal Year 1917 programme. What tipped the balance was the 'Preparedness' campaign launched in 1915, to ensure that the US Navy would be ready if Congress allowed the United States to become involved in the war in Europe. There was another agenda, too. The General Board was convinced that whichever side won the war would declare war on the United States to dominate its foreign markets, and sooner rather than later. The ultimate goal became a 'Navy Second to None', and that implied a balanced fleet.

At the end of July 1915 the Board put forward a tentative programme of four battleships, six battlecruisers and six scout cruisers. By October this had been refined to support the existing battlefleet: four battlecruisers, four fast scouts and ten destroyers. This led to a five-year programme to be completed in 1922: ten battleships, six battlecruisers and 40 destroyers. Then came news of the Battle of Jutland, and it was obvious that the battlecruiser had been overrated, and as a result the Senate approved a final programme of ten battleships, six battlecruisers and ten scout cruisers. It was telescoped into three years, Fiscal Years 1917, 1918 and 1919, and marked the final commitment to what became the Omaha (CL-4) class scouts.

Although funding had taken a long time to secure, design work had been in hand even longer. In May 1910 the Naval War College was asked to submit characteristics for a scout, 'restricted to the specific purpose of searching for the enemy', excluding the task of screening the battlefleet or attacking enemy screening cruisers. This ultra-narrow definition of scouting seems to envisage an orderly sea battle, a phenomenon that has never been encountered outside formal exercises. In fact the scouts would have two distinct roles. One meant operating compar-

atively close to the battlefleet to give it warning of the whereabouts of the enemy fleet. But they would also be required to report on the enemy fleet leaving its home base, a task which could result in scouts steaming great distances independently.

The Naval War College insisted that speed was the principal attribute required; the scout must be able to escape from more powerful enemy ships, in order to gather vital information and report back to the battlefleet. As it could not be faster than a destroyer it would need an anti-destroyer gun-battery as well as a medium-calibre battery to see off hostile scouts. After much discussion the General Board framed its own requirement and submitted to the Secretary of the Navy in October 1910. Sustained speed would be 26kts, with sufficient fuel (preferably oil) for a range of 8000 nautical miles at 12kts. Armament would be six to ten 5in 51-cal guns and two below-water torpedo tubes. They would be protected by an armoured deck, 1in on the flat and 2in on the slopes.

The ships would be specially equipped for their mission. These included a mast for spotter and lookout stations, a long-base rangefinder specially mounted to reduce vibration, and a powerful radio set. The Board asked the technical bureaux to consider the provision of space for a dirigible airship or an aircraft. In May 1914 Admiral Earle of the Bureau of Ordnance suggested a more austere design, displacing about 5200 tons and armed with six 6in guns and two underwater torpedo tubes, but speed raised to 28kts. Armour would be a partial 3in belt and a 2in deck over the machinery, and range would be reduced to 5000 nm. Provision would be made for a floatplane and catapult.

The Bureau of Construction & Repair objected that the proposed scout would be too small, and that provision of armour would push displacement up to 6500 tons. The Bureau also claimed that the underwater tubes would be impractical (confirmed by Royal Navy experience with similar provisions in some of its light cruisers), endurance was too low, and 3in armour would not keep out destroyer shells (not supported by Royal Navy wartime experience). This resulted in a revised specification: a maximum of 7500 tons, a speed of at least 28kts, eight or ten 6in guns, a 4in belt over the vitals, a 2in full-length deck, and an endurance of at least 6000 nm at 14kts. Provision was also made for four seaplanes and two above-water torpedo-tubes. When refined, the specification resulted in eight 6in guns, two anti-aircraft guns, twin torpedo

The USS Chester *was the namesake of the USN's first class of scout cruisers, all commissioned in 1908. These ships were very lightly armed, being built with 2-5in guns and 6-3in guns, on a hull displacing 3750 tons. Chester and her sister* Salem *were the first turbine-engined ships in the USN.*

Opposite: USS Raleigh *was commissioned in February 1924, and was present at Pearl Harbor on 7 December 1941.*

tubes on either beam, and facilities for at least two seaplanes. Protection would be a 4in belt extending from the forward magazine to the steering gear, and a 2in deck at waterline level.

Chief Constructor Watt had the task of turning all these ideas into a real warship. Despite a heavy workload he produced a preliminary design by the end of July 1914. He was sceptical about the value of a full-length protective deck, and found that it would be very heavy. Both ahead and broadside fire were important, so Watt mounted two 6in guns side by side on the forecastle, with two more recessed in ports below them and two more at the break of the forecastle. His design achieved astern fire by siting another pair of guns in tandem on the quarterdeck. The seaplanes could be hoisted in and out by midships derricks, which could also handle the boats. The drawback was cost; at $4 million the new scout would cost as much as four destroyers, and Watt suggested that expensive scouts might not be needed, 'in view of the increased size and seaworthiness of the recent destroyers…' Two world wars and half a century of intensive peacetime operations later, such faith in the flimsy and temperamental destroyers of the day seems misplaced.

A contrary voice was that of Admiral Frank Fletcher, Commander of the Atlantic Fleet. In a January 1915 letter to the Secretary of the Navy he pointed out that in the recent Naval War College fleet exercise, the Blue Fleet utterly failed in its mission because the destroyers could not operate in the heavy seas running. They were forced to reduce speed to 15kts and then 10kts, and never made contact with the Red Fleet. Fletcher said that the modern battleships were steaming at 19kts and even the old low freeboard ships were capable of 16kts. In his opinion there was a clear need for a 'heavy scout'.

There now ensued a period of bizarre proposals, including 12in gunned ships displacing over 14,000 tons, 14in gunned ships and even a 16in cruiser with virtually no protection, which only goes to show that other people besides the Royal Navy's Admiral Fisher were capable of producing monomaniacal concepts. These designs were influenced by Fisher's battlecruisers, and the original idea of a light cruiser capable of scouting seems somehow to have been subsumed into a battlecruiser concept. Inevitably, such a hybrid design fell between two stools, too big for cruiser work and too weak to be a fast capital ship.

By the autumn of 1915 these bizarre designs had been abandoned for what was now known as the 1917 scout. Speed was to exceed 30kts, endurance to be 10,000 nm and the battery was to be at least six 6in guns, all on a displacement of about 8000 tons. Attempts to return to a 12,000-ton design were voted down, and the General Board accepted an increase in armament to ten 6in and four 3in anti-aircraft guns, two above-water torpedo tubes and a 3in belt combined with a 1.5in deck. Secretary of the Navy Josephus Daniels approved these characteristics on 21 December 1915.

The new Chief Constructor, Rear Admiral David W Taylor, cannot have found the task of reconciling these requirements an easy one. Weight-saving had to be enforced ruthlessly:

The conning tower was deleted;
The sick bay was very small, with no operating theatre;
No baths, and no tiling in the crew's washrooms and elsewhere;
No deck planking;
Hammocks instead of bunks;
No mechanical ammunition hoists.

There were other economies, and an old-fashioned configuration of four slim funnels was chosen in mid-February 1916. Taylor provided a breakdown of weights in April, and contract

drawings were sealed on 8 July. In the light of the number of unusual features, representatives of the leading East Coast shipyards were invited to examine the drawings. The layout of armament struck an old-fashioned note, with sponsoned 6in guns forward of the bridge and abaft the mainmast, and a pair in the waist. While the ships were under construction the armament was augmented by the addition of two twin enclosed light turrets forward and aft, at the expense of the waist guns. Six more torpedo tubes were also added.

In 1917–18 the US Navy met the Royal Navy at close hand, and officers realized that the two navies' operational philosophy differed widely. Reports noted that the RN Centaur class could fire one more 6in gun on the broadside, despite displacing 2000 tons less. They also noted that British light cruisers carried all their torpedoes in their tubes, ready for immediate action. The observers were even more impressed by HMS *Raleigh*, whose six 7.5in guns completely outclassed the new scouts, although the US Navy ships would be the fastest in the world, 3kts faster than any Royal Navy cruisers built or building.

As a class, the Omahas served on routine peacetime duties, and had a relatively lucky Secnd World War, with no losses, although the *Raleigh* was damaged by a Japanese aerial torpedo at Pearl Harbor on 7 December 1941 and the *Marblehead* was damaged by air attack in the East Indies on 4 February 1942. The *Milwaukee* was lent to the Soviet Northern Fleet in April 1944 and renamed *Murmansk*. After considerable procrastination the Soviet Government finally returned the ship in 1949. She was the last to go to the breakers' yard, the others having gone three years earlier.

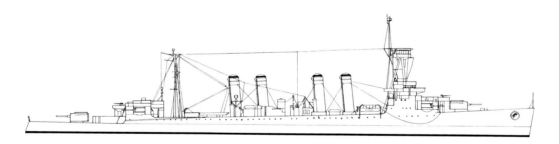

The Omaha class were old-fashioned in appearance and uncomfortable for the crew. (Line drawing 1/1250 scale.)

OMAHA CLASS

Omaha (CL-4) laid down 6 December 1918, launched 14 December 1920, commissioned 24 February 1923, built by Todd SB, Tacoma

Milwaukee (CL-5) laid down 13 December 1918, launched 24 March 1921, commissioned 20 June 1923, built by Todd SB, Tacoma

Cincinnati (CL-6) laid down 15 May 1920, launched 23 May 1921, commissioned 1 January 1924, built by Todd SB, Tacoma

Raleigh (CL-7) laid down 16 August 1920, launched 25 October 1922, commissioned 6 February 1924, built by Quincy

Detroit (CL-8) laid down 10 November 1920, launched 29 June 1922, commissioned 31 July 1923, built by Quincy

Richmond (CL-9) laid down 16 February 1920, launched 29 September 1921, commissioned 2 July 1923, built by Cramp, Philadelphia

Concord (CL-10) laid down 29 March 1920; launched 15 December 1921, commissioned 3 November 1923; built by Cramp, Philadelphia

Trenton (CL-11) laid down 18 August 1920, launched 16 April 1923, commissioned 19 April 1924, built by Cramp, Philadelphia
Marblehead (CL-12) laid down 4 August 1920, launched 9 October 1923, commissioned 8 September 1924, built by Cramp, Philadelphia
Memphis (CL-13) laid down 14 October 1920, launched 17 April 1924, commissioned 4 February 1925, built by Cramp, Philadelphia

Displacement:	7050 tons (standard), 9508 tons (full load)
Dimensions:	555ft 6in (oa), 550ft (wl) x 55ft 4.5in (max.) x 17ft (max)
Machinery:	4-shaft geared steam turbines; 12 Yarrow boilers; 90,000shp
Speed:	35kts (designed)
Armament:	12-6in 53-cal guns (2 x 2, 8 x 1), 2-3in AA (2 x1) 6-21in torpedo tubes (2 x 3)
Protection:	3in belt, 1.5in deck, 1.5-3in bulkheads
Range:	9000 nm @ 15kts (designed)
Complement:	29 officers, 429 enlisted personnel

CONCLUSION

The Omaha class was the product of much confused thinking, as their much-abbreviated design-history shows. Too much was attempted on the displacement, and as a result some features were exaggerated at the expense of balance. Despite being the fastest cruisers in the world their endurance was a disappointment. On such a small displacement there was no possibility of installing turbo-electric drive, as planned for the battlecruisers. Fast-running geared steam turbines were inherently extravagant on fuel at cruising speed, so there was no chance of achieving the planned 10,000 nm at 10kts. In service 6500 nautical miles was more realistic, using geared cruising turbines.

In fact only *Omaha*, *Milwaukee* and *Cincinnati* were given 15kt cruising turbines. The *Raleigh* and *Detroit* had no cruising turbines, but a Curtis combination plant permitted a cruising speed of 25kts. The *Richmond* and the later units could cruse at 20kts. In service the difference between the 'short radius' ships, *Omaha*, *Milwaukee*, *Cincinnati*, *Raleigh* and *Detroit* and the 'large radius' *Richmond*, *Concord*, *Trenton*, *Marblehead* and *Memphis* was so great that in 1928 the Commander-in-Chief of the Fleet suggested operating the class as two separate tactical divisions.

Although described as good seaboats, the alterations added 2ft to the draught, and they tended to heel deeply on the turn, flooding out the torpedo tubes. The gun positions were also washed out in a moderate sea. The designed complement of 330 grew to 425, resulting in cramped accommodation and poor sanitary provisions (one set of heads and washing facilities for 24 men). In 1924 the Bureau of Construction & Repair admitted that little could be done to make the ships more comfortable because of the sheer volume taken up by the machinery and the heat generated. In 1929 the Commander-in-Chief said they were too hot in the tropics and too cold in northern waters because of poor insulation. Their light hulls leaked from high-speed steaming, resulting in contaminated fuel tanks.

They spent little time as scouts, being used as leaders for destroyer squadrons. This was ironic, considering the anguished internal debate between 1910 and 1916 over the characteristics of the ideal scout. In part this was the outcome of too much theorizing and the pursuit of a fleet suited to the 'perfect' sea battle, something which never existed. Only the experience of a modern war could inject realism into the debate.

HSwMS Gotland
Hybrid Cruiser

Royal Swedish Navy 1925–1962

After the First World War the Royal Swedish Navy sought to replace the elderly cruiser *Fylgia* and the minelayer *Klas Fleming*, but could not obtain funding for two ships. Its designers came up with an unorthodox solution, the *Flyplankryssare* or hybrid cruiser–seaplane carrier–minelayer. The concept had much to offer, in theory at least, combining the gun—and torpedo-armament of a light cruiser with the strike and reconnaissance capability of a squadron of seaplanes, and being able to lay mines as well.

A naval committee was given the job of examining the proposal, and in December 1926 proposed a very small carrier displacing only 4500 tonnes, capable of launching 12 seaplanes but with no mean of recovering them. The Naval Construction Board noted that better seakeeping was required, as well as armament suitable for a ship leading the destroyer flotillas. The main armament was increased, three twin 152mm guns in turrets and catapults aft for launching eight seaplanes. Displacement rose to 5500 tonnes, and speed was to be 28kts.

Parliament authorised SEK16.5 million for the 'aircraft cruiser' on 13 May 1927 and the final plans were authorised on 31 March the following year. Unfortunately, the bids from leading Swedish shipbuilders showed that the ship could not be built within the funds allocated. This forced the Naval Construction Board to redesign the ship to fit within the budget limit. The most obvious change was the abandonment of the original layout of three twin 152mm L/55 Bofors Model 1930 gave way to a twin turret forward and another aft, and single 152mm guns in casemates, one on either beam, alongside the bridgework. The secondary armament comprised four 75mm M/1928 L/60 anti-aircraft (AA) and four 25mm M/1932 L/64 AA guns. 80–100 mines could be embarked, depending on type. Two sets of triple 533mm torpedo tubes were positioned on the weather deck, one on either side of the after funnel. They were controlled from the bridge.

The ship was only lightly armoured, with a 2-1.125in nickel-chrome steel deck, 1.125-6.25in transverse and longitudinal bulkheads, 1.125in on the funnel uptakes, 2-1.25in on the main turrets, 2-l.125 and 1-1.12in on ammunition hoists, and 2in protection for the conning tower. The armoured deck was an integral part of the transverse structure. The hull was flush-decked, with a raised aircraft platform running from the after gun turret to the stern. The mines were stowed on the weather deck, and discharged through ports over the stern.

The 152mm guns in the twin turrets had a maximum elevation of 60 degrees, and could in theory engage aircraft with variable-time (VT) fused shells. The two single casemate guns were however, restricted to 30 degrees and had limited arcs of fire. Range was 25,000m for the twin turret guns, but 5000m less for the single casemate guns.

The aircraft arrangements comprised a pair of sided Heinkel compressed-air catapults built by Deutsche Werke AG, a flush deck running to the stern, and a heavy-duty crane right aft. Eight seaplanes could be stowed on deck, and three more below it, but the *Gotland* never operated more than six seaplanes. The aircraft were export versions of the Hawker Osprey, a twin-float, folding-wing, two-seater biplane. Their engines were 640hp Pegasus air-cooled radials. Each seaplane had its own trolley to move the aircraft onto the catapult, and one could be launched every two minutes. The trolleys were self-propelled, with small electric motors, to move the seaplanes from the outer tracks to the centre track. The catapults were 14m in length when stowed fore and aft, but extended to 22m when in use.

For a ship of her displacement the *Gotland* had a relatively modest powerplant. Two sets of De Laval geared steam turbines and four French Penhoët boilers produced a moderate speed of 27.5kts at 300rpm, equivalent to 33,000 shp. On a high-speed trial in September 1934 she just exceeded her designed speed by 0.03kts and developed 32,768 shp. Getting it almost exactly right was a credit to the Navy's engineering branch; in some other navies the design-process was far less precise. The cruising range of 4000 nm was not a serious drawback, given that the ship's intended operational theatre was the Baltic.

After lengthy trials HSwMS *Gotland* was commissioned on 5 December 1934, and the following year began to operate with the Fleet to explore her potential. From December 1935 she began a series of foreign cruises, the last in June and July 1939 to Bordeaux and Southampton. When war broke out two month later she joined the Coastal Fleet, while continuing her training duties in Swedish waters.

HMSwS GOTLAND

Gotland laid down 1930, launched 14 September 1933, commissioned 5 December 1934, built by AB Götaverken, Göteborg

Displacement:	4750 tonnes (standard), 5550 tonnes (full load)
Dimensions:	442ft 3in (oa) x 50ft 7in x 18ft (at 4700 tonnes)
Machinery:	2-shaft de Laval geared steam turbines, 33,000 shp; 4 Penhoët boilers
Speed:	27.5kts
Armament:	6-152mm L/55 guns (2 x 2, 2 x 1), 4-75mm L/60-cal AA guns, 4-25mm L/64 AA guns,
Aircraft:	11 seaplanes (maximum stowage, but only 6 embarked)
Range:	4000 nm @ 12kts
Fuel:	800 tonnes (normal), 880 tonnes (maximum)
Complement:	467

The Gotland *attracted a lot of interest from foreign navies when she entered service in 1934, but technological developments rapidly rendered her obsolete. (Line drawing 1/1250 scaled.)*

The aircraft facilities proved a disappointment. In rough weather the seaplanes stowed on deck were very vulnerable to wave-damage, and the ship often had to return to harbour rather than risk further damage. In any case, the concept was very old-fashioned when compared with contemporary aircraft carriers in other navies. In 1943, therefore, the aircraft facilities were removed, and the former flight deck was extended forward (over the area formerly occupied by the catapults) and used to site four twin 40mm L/70 Bofors M/1936 AA guns. In addition two 20mm L/70 M/1940 guns were added. The reconstruction was finished in April 1944, and the Gotland rejoined the Coastal Fleet.

In the winter of 1946–47 the ship sailed on her first post-war foreign cruise, and in 1948 resumed the pattern of one annual foreign cruise during the winter and serving with the Coastal Fleet in the summer.

In 1953 she was taken in hand for a second reconstruction. This showed just how much naval warfare had changed since the design was first conceived nearly 30 years earlier. In addition to maintaining her peacetime role as a training ship, she was to have a wartime role as a fighter direction ship. Fighter direction was a concept developed during the Second World War, using a variety of hulls equipped with all the facilities of a shore-based fighter direction centre, including powerful long-range radars to track both friendly and hostile aircraft and vector the defenders to the point of interception.

To make room for new equipment the armament was altered. Out went the single 152mm casemate mountings and all the 75mm, 25mm and 20mm guns, as well as the 8mm machine guns. One twin 25mm mounting was retained, but shifted to the roof of the forward 152mm gun turret. In their place were installed five 40mm L/60 M/1948 single guns. Full radar-controlled fire was now possible, using Royal Navy-pattern 262 fire control and Type 293 surveillance radars. In addition she was fitted with an M/1952 automatic launcher for 103mm illuminant flares and a Type 144 sonar, also supplied by the Royal Navy. The original 4-metre rangefinder was replaced by more modern optical back-up fire control, and the three searchlights were removed. This final reconstruction was completed in 1954, and she rejoined the Fleet, but she had sailed on her last foreign cruise.

The Royal Swedish Navy had a long but unpublicized relationship with the Royal Navy, going back to the Russian War of 1854–56. Then it took the form of technical assistance as quid pro quo for Sweden having entering the war against the Russians as it neared its end. More technical assistance followed, as Sweden was seen as a buffer against a resurgent Russian Baltic Fleet; Sweden was given access to the Whitehead torpedo, for example, and was encouraged to work with British torpedo-boat builders. Thus the Cold War merely provided an excuse to continue a broadly unfriendly attitude to the Soviet Union. Evidence from British records shows that new radars and sonars were often 'lent' to the Swedes for evaluation, in return of the chance to evaluate Swedish ship—and equipment-designs.

By 1960 *Gotland* could not exceed 25kts, and was clearly worn out. There was little point in spending more money on her, and on 1 July that year she was removed from the Navy List. On 1 April 1962 she was sold for scrapping and was broken up the following year. Her career spanned 28 years, testimony to the fact that the Royal Swedish Navy took good care of its ships and did not wear them out by keeping them running year in and year out.

Conclusion

Although hailed as one of the most advanced warships of her day, the *Gotland* suffered the fate of most multi-role warships. She was far too small to be a successful seaplane carrier, and hardly fast enough to challenge the new generation of German light cruisers emerging in the early

1930s. As already noted, the concept was not adopted by any of the major navies, and however impressive she seemed to commentators at the time, carrier aviation in the rest of the world was making giant strides.

The Swedes fell into the trap of designing a ship to meet financial constraints, rather than working out what was needed militarily. Hybrid warships usually result in compromises which reduce their effectiveness in all roles. Although the *Gotland*'s hull was much too small, she would have been marginally more effective if she had been designed as a seaplane carrier pure and simple. Apart from a brief flirtation with heavy anti-ship guns in the US Navy and Imperial Japanese Navy's large carriers, the only valid carrier doctrine turned out to be that of treating the air group as the ship's main armament. Logically, therefore, all features which reduce the effectiveness of the air group must be eliminated or scaled-back as far as possible. When in the middle of the Second World War a group of naval aviators lobbied for a 'battleship-carrier' with triple 16in guns forward and aft and a flight deck in between, a senior flag officer commented that it was the product of 'gross psychological maladjustment'.

The provision of two extra 152mm guns with restricted range and arcs of fire is very revealing to a modern analyst. On such restricted dimensions one might have thought that the weight could be put to better use elsewhere in the ship. The Baltic can be stormy in winter, and casemate positions for guns in much larger warships in the British, German and US navies had proved very liable to be flooded in heavy weather. We can only conclude that the Royal Swedish Navy designers did not take such considerations very seriously, one of the consequences of long isolation and a mindset that subconsciously, or even consciously, discounted the risk of ever going to war.

Another flaw in the concept was the assumption that the seaplane was adequate to meet the requirements of what is known today as the strike mission. The Osprey was neither better nor worse than its contemporaries. Like them it was fragile and ran the risk of sustaining serious damage when stowed on deck in bad weather, and might all too easily be impossible to hoist back on board. Reconnaissance was within the Osprey's capabilities, but not combat against hostile fighters. The Royal Swedish Navy was not alone in overrating the capabilities of the seaplane, which could never match the performance of wheeled aircraft because too much weight was taken up by the twin floats.

In response to the question of what would have been the correct dimensions and displacement for the ship, the answer is not obvious. In the closing months of the First World War the Royal Navy had converted the hull of a 9750-ton cruiser, HMS *Vindictive*, to a hybrid carrier-cruiser, with a flight deck aft and an armament of single 7.5in (190mm) guns. She served in the Baltic in the Intervention War in 1919–20 but was regarded as too small, an accusation levelled at her similar-sized contemporary, the first Royal Navy ship built as a carrier from-the-keel up, HMS *Hermes*.

Perhaps the best option would have been a larger light cruiser, displacing about 7000 tonnes and capable of launching one or two seaplanes for reconnaissance and spotting for the guns. But that is an opinion based on hindsight; clearly the Swedish Government was keen to keep size down and save money. In the final analysis a navy gets the type of ship which it thinks it wants at the time. Designers can only implement the official policy, no matter how much they might wish to do otherwise.

Duquesne class Heavy cruisers

French Navy 1924–1962

Although the *Marine Nationale* had grumbled about the ratio of tonnage allowed to the signatories of the Washington Naval Treaty of 1922 (neither France nor Italy nor Japan had parity with the United States and Great Britain), they could not ignore the category of 10,000-ton, 8in-gunned heavy cruiser. France had a large colonial empire scattered across the Americas, Africa, and East Asia, and a large mercantile marine. These would suffer if hostile cruisers were able to operate with impunity.

In 1924 two new 'Washington' cruisers were authorised as part of that year's programme, and ordered on 1 July the same year. Their design was clearly based on the 155mm-gunned *Duguay-Trouin* design laid down two years earlier, if for no other reason than the *Marine Nationale*'s lack of any other modern cruisers on which to base their ideas. They were to bear the names *Duquesne* and *Tourville*, commemorating famous admirals of the sailing era.

The Navy had no 8in (203mm) gun so the 203mm 50-call Modèle 1924 gun and its mounting had to be designed from scratch. The projectile weighed 127kg, fired at a muzzle velocity of 850m per second over a maximum range of 31,000m. The 75mm Modèle 1922 were the same as those which armed the light cruisers, but double in number, grouped amidships and around the after control position. They also received four of the new 37mm Modèle 1925 single anti-aircraft guns, backed up by 13.2mm machine guns. The Modèle 1923 DT 550mm torpedoes had a range of 20,000m at 29kts, or 10,000m at 35kts. The warhead weighed 415kg.

The French Navy's designers soon found, as had everyone else, that a balanced design for a ship displacing no more than 10,000 tons, armed with a reasonable number of 203mm guns and adequately protected, required more than the normal amount of compromise. 'Adequate' protection in this context was usually considered to be a combination of deck and side armour capable of resisting the fire of similarly armed enemies. On that score the Duquesne class failed, as they had only 30mm box citadels protecting the main magazines, and the same thickness on the deck, turrets and conning tower. A more modest steam plant might have freed some tonnage for thicker armour, but as it stood the 430 tons accounted for just 4.3 per cent of the total.

The machinery benefited from being arranged on the unit system, with machinery and boilers grouped separately. The boilers worked at slightly higher temperatures than those in the light cruisers, and both cruisers exceeded their designed speed on trials. *Duquesne* reached 35.3kts on trials, with 131,800 shp, but was beaten by *Tourville*, which reached 36.15kts, despite a lower output of 126,900 shp. Like their Italian rivals, these speeds were never attainable in service, and

The Duquesne *was flimsily built, and was worn out within twenty years.*

required some 'creative accounting' for weights; but both ships were economical steamers, capable of maintaining 30kts indefinitely at half-power.

The light anti-aircraft armament was not satisfactory. The 37mm gun was a semi-automatic weapon with too slow a rate of fire, while the 13.2mm 'heavy' machine gun lacked stopping power, even against the aircraft of the day. Later the 37mm were replaced by twin Modèle 1933 37mm, with a rate of fire of 85 rounds per minute (compared to only 30 per minute for the older gun).

Two GL-812 floatplanes were embarked on completion in 1928, with a single catapult sited on the centreline between the after funnel and the mainmast, but no hangar was provided. During their peacetime careers the floatplanes were replaced, first by GL-832s and later by a pair of Loire-Nieuport 130s, cumbersome twin-engined flying boats.

The navy was very unimpressed by these ships on account of their minimal protection, and in 1935 a serious proposal was put forward to convert them to light aircraft carriers. In this guise they would have embarked a theoretical air group of 12–14 aircraft each, but the proposal was vetoed in favour of a more coherent plan to build two large fleet carriers, the Joffre class.

In service both ships made long cruises and carried out normal peacetime missions. The *Duquesne*'s first public appearance was at a large Naval Review held at le Havre in July 1928, following which she set off on a long cruise to Guadeloupe via New York. In 1929 she circumnavigated Africa during a seven-month cruise, returning to Toulon to join the 1st Light Division as flagship, part of the 1st Squadron. In October 1931 she visited the United States to represent France at the 150th Anniversary of the British surrender of Yorktown. From 1932 to 1938 she was based at Toulon once more, serving with the 1st and 3rd Light Divisions, and in 1938 was attached to the Gunnery School as a training ship.

In 25 January 1940, nearly four months after the outbreak of war, she sailed for Dakar, where she joined one of the groups hunting for German commerce-raiders. In May she sailed for Alexandria as flagship of the 2nd Cruiser Squadron of Force X, formed to reinforce the Royal Navy in the Eastern Mediterranean. Her only major movement was an abortive sortie into the Adriatic on 12–13 June, after which she was immobilized at Alexandria on the orders of Admiral Godfroy. There she remained until late in June 1943, when she sailed for Dakar via the Cape of Good Hope, *en route* to the United States for a long overdue refit. She returned to Dakar, where she was a unit of the 1st Cruiser Division until December 1943, and remained there until April the following year. In May 1944 she arrived in the Firth of Clyde, to serve as a support ship during the Normandy landings, and was attached to *Groupe Lorraine* of the French naval task force. She bombarded German coast defence fortifications at Royan and Pointe de Grave. She was now in need of another refit, and was sent to Brest for a refit which lasted from June to November 1945.

The *Tourville* began her active career with a world cruise in April 1929, and did not return to Lorient until the following December. In 1930 she joined the 1st Light Division at Toulon, transferring to the 3rd Light Division in 1934. She helped to evacuate refugees during the Spanish Civil War and helped to enforce the League of Nations' non-belligerency policy between August 1936 and May 1937. She started a long refit at Toulon in January 1939, and did not return to active service with the 2nd Cruiser Division until August that year.

She was employed in searching for German merchantmen in the Mediterranean and sailed from Bizerte to Beirut in December 1939, stopping and inspecting 32 ships. Another trip was made from Toulon to Beirut between 20 January 1940 and 7 February, carrying a large gold shipment. She joined her sister in Force X at Alexandria, until 'demobilised' in July. On her return to service she formed part of the 1st Cruiser Division at Dakar. She was sent to Bizerte for a refit in June 1944, and then to Toulon in November, where she was used as a base ship for escort vessels.

Wartime modifications were few. After the two ships rejoined the Allies in 1943 the French light anti-aircraft armament was replaced by four twin 40mm Bofors guns and 16 single 20mm Oerlikon guns. The 40mm guns were grouped on a platform aft, available after the removal of the mainmast. A December 1945 photograph of *Duquesne* shows that US Navy-pattern radars were added. The torpedo tubes and catapults had been removed in 1943. By 1945 the average full load displacement had risen to 13,500 tons, presumably causing a noticeable reduction to maximum speed.

After the war both ships were involved in the ill-fated effort to suppress the Vietminh insurgency in Indochina. The *Duquesne* participated in two campaigns, one from January to November 1946 and another from December 1946 to August 1947, including shore bombardments and amphibious landings. She was paid off in 1950. She was put into Special Reserve Category 'A' and later became an accommodation ship at Arzew in Algeria until 'condemned' for disposal in 1955. The *Tourville* also did two tours of Indochina duty, the first from January to July 1946 and the second from October 1946 to November 1947. After her return to Toulon she was paid off about 1950. She became an accommodation hulk at Brest and eventually a mooring pontoon, until stricken in 1961. She was finally scrapped in 1963.

DUQUESNE CLASS

Duquesne laid down 30 October 1924, launched 17 December 1925, completed 6 December 1928, built by Brest Naval Arsenal
Tourville laid down 4 March 1925, launched 24 August 1926, completed 1 December 1928, built by Lorient Naval Arsenal

Displacement:	10,160 tons (standard), 11,640 tons (normal), 12,395 tons (full load)
Dimensions:	185m (pp), 191m (oa) x 19m x 6.32m (mean)
Machinery:	4-shaft Rateau-Bretagne single-reduction geared steam turbines, 120,000 shp; nine Guyot du Temple boilers
Speed:	33.75kts

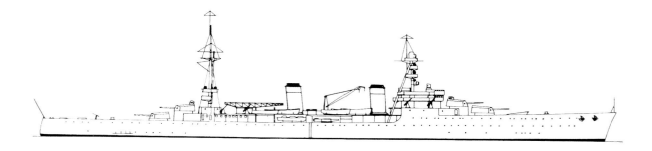

Armament:	8-203mm 50 cal. M1924 guns (4 x 2), 8-75mm M1922 (8 x 1), 4-37mm M1925 (4 x 1), 12-13.2 MGs (8 x 1), 12-550mm torpedo tubes (4 x 3)
Aircraft:	2 floatplanes
Armour:	30mm box citadel over magazines, 30mm deck, turrets and CT
Fuel:	1820 tons oil
Range:	4500 nm @ 15kts, 2000 nm @ 30kts
Complement:	30 officers, 575 ratings

The Duquesne class were an unsatisfactory compromise to meet the limits imposed by the 1922 Washington Naval Treaty. One hopes their failures contributed to the eventual success of the Algérie, France's, and one of the world's, best heavy cruiser built between the two wars. (Line drawing 1/1250 scale.)

CONCLUSION

Although the French Navy was not the only one accused of building 'tinclad' heavy cruisers to meet the requirements of the Washington Treaty, it can be argued that the Duquesne class were the worst. They were unfit for combat with any existing cruiser, because of their flimsy construction, despite their heavy main armament. Their endurance was hardly adequate for cruisers intended to protect commerce in the distant oceans, and showed that Mediterranean operations had precedence.

A more balanced design might have been achieved by trading off high speed against fuel capacity, but little could have been done to increase armour. No other navy solved this conundrum in the first generation of 'Washington' heavy cruisers, apart from the Japanese, who simply failed to declare any excess tonnage which contravened the Treaty.

The French designers faced a host of problems. They had not completed so much as one light cruiser since 1896. The designers and the active fleet had therefore missed out on the ultra-rapid developments of technology which had made possible the robust and battleworthy light cruisers of the pre-1914 years. Nor did the French have the benefit of these ships' extensive war experience in the North Sea and the Adriatic. As part of the spoils of war the French Navy had received five ex-German and ex-Austro-Hungarian light cruisers, including the *Mulhouse* (ex-German *Straslund*), *Colmar* (ex-German *Kolberg*), *Strasbourg* (ex-German *Regensburg*), *Metz* (ex-*Königsburg*) and *Thionville* (ex-Austro-Hungarian *Novara*). Such ex-enemy prizes rate highly in the public mind as tangible proof of recent victory, but they must have been a logistical nightmare to the post-Armistice French Navy. Keeping them running must inevitably have occupied dockyard resources to the maximum, and the technical lessons learned would be almost meaningless. All three Duguay-Trouin class ships were still in builders' hands when the Duquesne class were laid down, so any lessons to be learned from them were not yet available.

The French approach to the Washington Treaty was quite hostile, and the Navy's motive for eventually building heavy cruisers seems to have been a need to compete as an alternative to being outclassed. Contemporary French strategic studies show that the British were seen as the main enemy, rather than the Italians. This perception was to change with the expansion of Mussolini's military ambitions and the emergence of Nazi Germany, but it would explain the apparent lack of interest in traditional cruiser warfare in distant oceans. Although they were economical steamers, as already noted, their endurance at 30kts was a miserable 2000 nm. France was also obsessed with safeguarding its communications with its colonies in North and West Africa, both of which were expected to provide troops to defend France in the event of a European war. The troop reinforcements from North Africa had saved the nation in August 1914, and were expected to do the same again if needed. The inconvenient fact of the empire in the Far East being given a lower priority seems to have been largely ignored, although this may reflect the opinion that France's naval forces were not adequate to defend both, a dilemma which troubled the much larger Royal Navy.

The two ships were lucky in one sense; they did not face an opponent in battle, a situation in which their almost non-existent protection could have been suicidal. On the other hand, the catastrophic defeat in 1940 was followed by the 'regrettable action' at Mers el Kebir, in which the Royal Navy disabled or neutralized the major units of the fleet. Mers el Kebir split *La Royale* into Free French and pro-Vichy factions. In that period of divided loyalties ships like the Duquesne and Tourville never had a chance to show what they could do. Only the German decision to attack Toulon in 1942 healed the wounds, at the cost of virtually the entire Toulon fleet, scuttled to keep the ships out of German hands.

The drawback to building very light ships is their short lives. Although the two Duquesne class averaged only fifteen years of service each, with an enforced period of idleness in 1940–43, they were worn out by the end of the war and were not worth any serious post-1945 modernization. Other navies' cruisers were also worn out by 1945, but they had the excuse of continuous service and hard driving in all weathers, and in many cases battle damage as well. The cost of refurbishing such ships was so high that virtually all were laid up in reserve or relegated to subsidiary duties such as training.

France's naval industry had been in steady decline since the first years of the century, as attention turned from a naval conflict with the British to a confrontation with Germany on land. The First World War accelerated the decline; shipyard workers were conscripted into the army and some parts of naval industry were turned over to the manufacture of land ordnance. Reversing the decline would have taxed any government, but by 1918 France was exhausted by its huge loss of life, the destruction of its northern industrial heartland and the burden of war-debt. It is hardly surprising to find evidence of poor designs, but things did get better. The four Suffren class heavy cruisers were better protected than the Duquesne class, and the last 'Washington' heavy cruiser, the *Algérie*, was arguably the best of the type anywhere, well protected, fast, well armed and within the 10,000-ton limit. Sadly, she was scuttled at Toulon in November 1942.

Deutschland class 'pocket battleships'

German Navy 1927–1945

The Treaty of Versailles, imposed on a defeated Germany in 1919, bore particularly heavily on the former Imperial Navy. The rebuilt *Reichsmarine* (national fleet) was allowed to possess only six 'coast defence armoured ships' or *panzerschiffe* and six light cruisers. These were in fact obsolete vessels, superseded by more modern designs in every possible opponent, pre-dreadnought battleships and small light cruisers of pre-1914 vintage. The Treaty recognised the fact that they would eventually wear out, and 20 years was the service-life permitted; the major ship-replacements must not exceed 10,000 tons, and light cruisers were limited to 6000 tons.

Curiously, the Treaty did not specify the means of calculating displacement tonnage, nor was there any stipulated upper limit of gun-calibre. The latter question was left to a Conference of Allied Ambassadors, which permitted the gun-calibre of the new training cruiser *Emden* to be increased from 10.5cm to 15cm. The *Reichsmarine* opted, however, to stick to 28cm for the eventual *panzerschiff* replacements, as anything bigger would risk provoking the Allies. Retaining 28cm-calibre guns also eased the problems of the designers, who would have been hard put to get 30cm or 38cm guns into a design intended to displace only 10,000 tons in 'standard' condition.

In the 1920s the *Reichsmarine* envisaged a role for the future *panzerschiffe*: a French attack in the Baltic in support of its treaty obligation to Poland, at least a cruiser squadron, probably reinforced by a one of the surviving 'semi-dreadnoughts' dating from before the war. Contrary to what has previously been written about these ships, they were not at this stage intended for commerce-raiding. The *Reichsmarine* was too weak to take any risks which would involve it in a confrontation with France's large navy.

In 1923 work started on the first new designs, but it was suspended the following year when the economy collapsed. The new head of the Navy, Admiral Wolfgang Zenker, was anxious to get the process restarted, and in 1925 the team prepared three new sketch designs, making five in all:

I/10	a 32-kt cruiser armed with eight 20.5cm guns
II/10	a 22-kt armoured ship armed with four 38cm guns
II/30	armed with six 30cm guns and protected by 20cm armour
IV/30	a more lightly armoured version of II/30
V/30	with six 30cm guns and 18cm armour

It is interesting to note that the last three designs were diesel-powered to save weight.

A meeting held in May 1925 to discuss the designs was inconclusive, which is hardly surprising, as the Construction Office was fully committed to work on a new type of light cruiser and a replacement for worn-out torpedo boats. Nor could new heavy gun mountings be built quickly or easily, as the French still occupied the Ruhr. But in spite of the workload the Construction Office produced a new series of 28cm-gunned designs:

I/35   a slow but well-armoured ship with a triple turret forward
VIII/30   with lighter protection and two twin turrets, one forward and the other aft

Only when the French withdrew from the Ruhr in the following year could the *Reichsmarine* begin to think seriously about the construction of heavily gunned ships. It had been hoped to lay down *Panzerschiff* 'A' in 1926, but a final design had not yet been selected and the shipyards were demanding work of any kind, so another light cruiser and a flotilla of torpedo boats were ordered that year. The delay was beneficial, however, for it gave time to absorb the lessons learned in the 1926 Fleet Manoeuvres. The principal 'discovery' was the obvious one that higher speed conveyed great tactical advantages.

Two new designs were offered, of which Admiral Zenker chose 'C' because he liked the combination of 28cm guns and the relatively high speed of 26kts. The use of diesels would be a further advantage, giving the new ship long range for ocean warfare. It also won the support of Vice Admiral Erich Raeder, Commander of the Baltic Fleet, whose ideas on commerce-raiding were beginning to take shape. The designation for the new design was *panzerschiff* (armoured ship) to avoid any suggestion that a battleship was planned.

The *Reichstag* voted the first credit for the new ship in 1927; she was designated *Ersatz Preussen* (*Preussen* replacement), as required by law. There was considerable opposition to the new ship, and it was decided to defer the order until the election for a new *Reichstag* took place in May 1928. Even then, it was November 1928 before final approval was given. The Socialists campaigned under the banner of 'Battleships or Child Welfare?', but 810,000 Germans voted for Hitler's National Socialist Party and won the Nazis 12 seats in the *Reichstag*. Their votes helped to secure political approval for the new *panzerschiff*, and the Navy was to show its gratitude. At the launch of the new ship Chancellor Brüning gave voice to the symbolism: '…witnessed by the entire world, the German people have demonstrated that, despite the shackles imposed upon them and all their economic problems, they have the strength to guard their peaceful co-existence with the rest of Europe'. The name chosen was *Deutschland*, bestowed by President Paul von Hindenburg.

When Hitler came to power in 1933 the rebuilding of the *Reichsmarine* was proceeding well, with the *Deutschland* close to completion. The second *panzerschiff*, *Admiral Scheer*, would be in service before the end of 1934, and the third, *Admiral Graf Spee* was to be ready early in 1936. A fourth had already been approved by the *Reichstag* and Hitler approved the construction of a fifth, *Schlachtschiff 'E'*. However, it was to be kept secret because Raeder intended to build a pair of 26,000-ton ships. This was the first intimation that the *panzerschiff* programme was to be cut short; the fourth unit was ordered as *Schlachtschiff 'D'*, and no.5 became *Schlachtschiff 'E'*. These ships became the *Scharnhorst* and *Gneisenau* respectively, and were to be armed with three triple 28cm turrets each, ordered as long-lead items for *panzerschiffe* 4, 5 and 6 (the sixth was never laid down).

Meanwhile the formidable Nazi propaganda machine made much of the *Deutschland*'s technical innovations, which allowed the design to remain inside the 10,000 tons permitted by the

The Admiral Graf Spee *at the 1936 Coronation Naval Review.*

Versailles Treaty. Naturally, no mention was made of the fact that her true standard (Treaty) displacement was 11,700 tons, a 17 per cent excess. They were the first large warships to be driven by diesel engines, and to use large-scale electric welding, but all three of the class suffered continually from machinery problems and vibration. The foreign press swallowed all the false claims, and the British popular press boosted German prestige even more by labelling them 'pocket battleships'. According to many analysts, only seven warships in the world were capable of dealing with a single 'pocket battleship': the three Royal Navy battlecruisers and the Japanese *Kongo* class.

When Hitler announced in a speech on 5 November 1936 that Germany's enemies were Great Britain and France, Raeder realised that his navy, now renamed the *Kriegsmarine* (war fleet) as a symbolic gesture of defiance to the Versailles Treaty system, would sooner or later face the might of the Royal Navy. His answer was to develop a commerce-raiding strategy which would use the growing surface fleet to disrupt the convoy system and allow the U-boats to destroy merchant ships at will. The success of the convoy system in defeating the U-boats in 1917–18 and the impotence of the High Seas Fleet to affect the outcome was burned into the collective psyche of the German Navy's officer corps. There was also the disgrace of the scuttling of the High Seas Fleet in Scapa Flow in 1919, a stain which the officer corps felt could only be erased

The hulk of the scuttled Admiral Graf Spee *burns in the River Plate esturary in December 1939.*

by defeating the Royal Navy in battle.

When war finally came in September 1939 the *Kriegsmarine* was ready. The *Admiral Graf Spee* had sailed from Wilhelmshaven on 21 August, followed three days later by the *Deutschland*, and two supply ships were also sent out into the Atlantic to support them. Bad weather helped the ships to avoid being spotted by Royal Navy patrols, and on 28 August the *Deutschland* and her supply ship, the *Westerwald*, passed through the Denmark Strait and entered the North Atlantic. The *Admiral Graf Spee*, with her support ship *Altmark* had passed east of Iceland three days earlier, and both ships had reached their war stations before the first *Wehrmacht* troops had crossed the Polish border.

Both ships started to sink or capture victims, but Hitler was obsessed by the effect on German morale if a ship called *Deutschland* was sunk, and she was recalled, berthing at Gotenhafen (Gdynia) on 15 November. She was immediately renamed *Lützow* on a personal order from Hitler. Unlike her half-sister, her cruise had yielded only meagre results, but she had tied down a number of British and French warships.

The successful commerce-raiding career of the *Admiral Graf Spee*, and her scuttling off Montevideo on 13 December 1939 have been told and retold many times, and are not relevant to this study. What is very significant, however, is the reason for her final destruction. Her diesel engines were giving trouble by the middle of November, and although her engineers had done some repair work, Captain Langsdorff radioed a detailed account of the poor state of his ship's engines to Berlin. Her luck was beginning to run out, and Langsdorff knew that some of his victims' reports had reached the Admiralty. The *Graf Spee*'s Arado floatplane was vital, both as a scout to detect enemy merchant ships and as a means of avoiding hostile warships. It became inoperable on 11 December, when a cylinder block cracked.

At 05:52 on 13 December the German ship sighted the masts of the heavy cruiser HMS

Exeter, and at 06:16 learned that she was accompanied by the light cruisers HMS *Ajax* and HMNZS *Achilles*. Langsdorff felt confident that he could defeat the cruisers, but Commodore Henry Harwood RN had very clear instructions on how to handle his ships. By manoeuvring independently his three ships would provide too complex a set of problems for the *Admiral Graf Spee's* fire control. HMS *Exeter* with her 8in guns was the most powerful British unit, and the two small cruisers were disposed well clear of her to provide 'flank marking' to improve the *Exeter's* shooting. Langsdorff knew that HMS *Exeter* presented the biggest threat, and ordered fire to be concentrated on her; the British cruiser suffered severely but the *Ajax* and *Achilles* closed to draw fire away from her. The *panzerschiff* did not escape unscathed, as splinters from near-misses holed her below the waterline and two 8in shells hit. Lansgdorff also suffered a small injury from a shell splinter. The 6–7kt speed advantage of the three cruisers also helped them to avoid destruction. Finally the frustrated Langsdorff ordered a turn to the west and made his escape at about 24kts, leaving the British to deal with the crippled *Exeter*. His ship had expended 60 per cent of her 28cm ammunition, and he took her into neutral Montevideo to seek a respite, bury the dead and try to repair the damage.

A complicated diplomatic wrangle ensued as the British tried to delay her departure until reinforcements could arrive off the River Plate. What the British did not know was that Langsdorff sent a pessimistic message to Berlin, stressing the poor state of the *Admiral Graf Spee*. In particular he stressed a hole in the forecastle which would endanger the ship when she reached the North Atlantic. After long deliberations the Naval Staff instructed Langsdorff to fight his way out or scuttle his ship, but on no account to submit to internment. He chose to scuttle her in Uruguayan territorial waters and shortly afterwards shot himself, probably in a severe state of depression.

The Admiral Scheer *as originally built in 1935. She would later receive a new funnel cap and a rakish bow. (Line drawing 1/1250 scale.)*

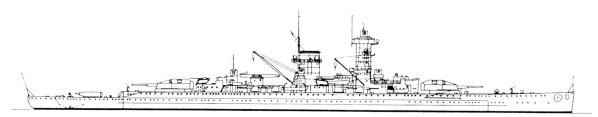

DEUTSCHLAND CLASS

Deutschland laid down 5 February 1929, launched 19 May 1931, commissioned 1 April 1933, built by Deutsche-Werke, Kiel

Admiral Scheer laid down 26 June 1931, launched 1 April 1933, commissioned 12 November 1934, built by Wilhelmshaven Navy Yard.

Admiral Graf Spee laid down 1 October 1932, launched 30 June 1934, commissioned 6 January 1936, built by Wilhelmshaven Navy Yard.

DEUTSCHLAND AND ADMIRAL SCHEER

Displacement:	11,700 tons (standard), 15,900 tons (full load)
Dimensions:	186m (oa) x 21.6m x 7.4m (maximum)
Machinery:	3-shaft MAN diesels, 48,390 bhp (*Scheer* 52,050 bhp)
Speed:	26kts
Armour:	3ft 9in belt, 1.5in deck, 5ft 6in CT and turret faces, 4in turret sides and barbettes.

Armament:	6-28cm (2 x 3), 8–15cm (8 x 1), 6-105mm AA (6 x 1), 8-21in TT (2 x 4)
Aircraft:	2 floatplanes
Complement	1150

ADMIRAL GRAF SPEE

Displacement:	12,100 tons (standard), 16,200 tons (full load)
Dimensions:	186m (oa) x 21.6m x 7.4m (maximum)
Machinery:	3-shaft MAN diesels, 54,000 bhp
Speed:	26kts
Armour:	3ft 9in belt, 1.5in deck, 5ft 6in CT and turret faces, 3in turret sides, 5in barbettes.
Armament:	6-28cm (2 x 3), 8–15cm (8 x 1), 6-105mm AA (6 x 1), 8-21in TT (2 x 4)
Aircraft:	2 floatplanes
Complement	1124

CONCLUSION

The fate of the *Admiral Graf Spee* and the meagre successes of her sisters *Admiral Scheer* and *Lützow* underline the flaws in Raeder's strategy. Despite their propaganda value and the semi-hysterical reaction of the British popular press, they were very expensive, over-gunned heavy cruisers, with only a moderate turn of speed and a light-cruiser scale of protection. They were a very expensive solution to the problem of tackling the British Merchant Navy, and it is significant that the later raiders, converted merchant ships, were far more successful. When used as quasi-capital ships they were failures; In December 1942 the *Lützow* and the heavy cruiser *Admiral Hipper* were driven off by a convoy's escorting destroyers.

The Admiralty's well-rehearsed tactics of splitting a *panzerschiff*'s fire control worked very well at the Battle of the River Plate; earlier in 1939 the *Exeter* had played the role of a *panzerschiff* in an exercise. The DNC, Sir Stanley Goodall, noted in his diary after the scuttling of the *Admiral Graf Spee*, 'I always said that they [the *panzerschiffe*] could be handled by a couple of "Counties" [better-armed than the *Exeter*]'.

The design was well-executed, with many innovative features, but the true standard displacement of 11,700 tons shows that German ingenuity could not bypass the laws of physics and hydrodynamics. The speed requirement proved inadequate because it had in mind the contemporary speed of battleships, 23kts. By 1939 a new generation of fast battleships had appeared, so the claimed ability of a *panzerschiff* to avoid action proved illusory. In any case the contemporary 'Treaty' heavy cruisers and light cruisers had an even bigger margin of speed. The *panzerschiff* as a showcase for German technology justified its creation, but was of limited value as a battle-unit. It came as a rude shock to the *Kriegsmarine* to realise that the ships were so vulnerable to 8in gunfire.

The other lesson which had to be re-learned was the way in which comparatively minor damage could cripple a commerce-raider. The long wrangle with the Uruguayan Government was inevitable, in the absence of a friendly base. Any action against a warship carried with it the risk of such damage, and heavy expenditure of ammunition. With vastly improved communications (most merchant ships now had radios) the feats of the commerce-raiders in the First World War were never likely to be repeated. As Admiral Mahan had predicted, commerce-raiding could inflict losses, but it would never defeat a powerful navy.

CONDOTTIERI CLASS LIGHT CRUISERS

ITALIAN NAVY 1928–1972

Italy had fought on the Allies' side for most of the First World War. In 1922, however, Italian politics changed radically under the impact of social unrest on the part of workers and veterans of the war. Fear of civil war caused the king to give the demagogue Benito Mussolini, leader of the Fascist movement based on an alliance between wealthy Italian industrialists and landowers and aggrieved veterans, the authority to form a government. Mussolini believed Italy was best

Two Condottieri class cruisers in drydock. The class worked hard from the start of the war, but their vulnerabilities resulted in several sinkings.

The Bartolomeo Colleoni *passes through the Suez Canal in pre-war years.*

served by a nationalist foreign policy and the country's rival in recent decades in the Mediterranean had been France. Consequently the Navy started to plan for a possible war with France.

The construction by the French Navy of a new class of 2100-ton *contre-torpilleurs* or large destroyers, armed with 130mm guns in the early 1920s forced the Italian Navy to seek some form of antidote. The first efforts included the building of *esploratori* or scouts, the *Navigatori* class, and a series of larger scout cruisers, in the period 1928–31. The latter were named after famous (or infamous) *condottieri*—the mercenary generals who had dominated Italian warfare in the 14th and 15th centuries.

The scout cruisers' task was to catch and destroy the very fast French *contre-torpilleurs*, so high speed was given top priority. In fact virtually no protection was provided, and the survival of the ship was based on superior gunpower. To comply with the Washington Treaty they were rated as light cruisers, armed with 6in guns.

The construction of the hull used a mixture of transverse and longitudinal framing, with a degree of tumblehome. A raised forecastle extended for about a third of the length, providing good seakeeping. To protect the machinery spaces 25mm nickel-chrome steel side plating was provided, thinned to 20mm in way of the magazines forward and aft. The upper deck was 20mm thick, and the citadel was closed by 20mm bulkheads. An internal longitudinal bulkhead of 18mm splinterproof steel was provided. On a standard displacement of 5191 tons, 584 tons was allocated to protection, i.e. only 11.3–11.5 per cent of the total.

The machinery consisted of two-shaft Belluzzo geared steam turbines, developing a designed maximum of 95,000 shp for a sea speed of 30kts at full load. Steam was generated by six Yarrow-Ansaldo boilers, with the two forward pairs of boilers serving the starboard turbines, and the after pair serving the port turbines. Unfortunately, the government had the idea of offering a substantial bonus to the builder for every knot over the designed speed. It led to forcing of machinery during the builder's trials to produce misleading speeds. *Alberico da Barbiano*'s turbines developed 123,479 shp, and reaching 42kts in the lightest possible condition. This would, it was hoped, stimulate interest from export customers in Italian warships, but in practice it meant that the ship would never get anywhere near that speed again.

The main armament comprised a rational layout of four twin 152.4mm (6 gun turrets, two forward and two aft. The Ansaldo 152.4mm 53-cal Model 1926 had the high muzzle-velocity of 900m/second, and the arrangement of two guns in a single sleeve, close together, made for a less than satisfactory mounting. The projectile weighed 47.5kg, and the combination of a light shell and high velocity reduced accuracy at the longer ranges. The secondary armament of three twin 100mm 47-cal Odero-Terni-Orlando Model 1927 mountings was disposed on the beam amidships (two) and a third on the centreline. In common with many navies, the demands of air defence were met by the provision of four light anti-aircraft mountings: four twin 37mm 54-cal, two on the after end of the hangar roof and two aft. Breda 13.2mm machine-guns made up the rest of the light armament. All except the *Alberto di Giussano* could be fitted with mine rails on the quarterdeck. All had two twin 533mm torpedo tubes, positioned abreast of the after funnel.

With typical Italian ingenuity, the requirement to operate two Cant 25 AR floatplanes was met by the provision of a small hangar incorporated into the forward superstructure. The Magaldi powder catapult ran from 6in turret 'A' along the full length of the forecastle. Aircraft were run out through side doors and then on tracks to the catapult. Of all the expedients adopted between the two world wars to operate reconnaissance floatplanes, this was the least successful, guaranteed to expose the aircraft to weather damage, or limiting operations to calm weather.

Experience with the first-of-class, the *Alberto di Giussano*, showed that she lacked an adequate margin of stability, so the tripod mainmast was replaced by a light pole mast. The hull also proved to be too light for driving at high speed, and had to be stiffened, reducing speed. In 1938–39 the 37mm guns were replaced by Breda 20mm 65-cal.

The ships' active service included involvement in the Abyssinian Crisis in 1936 and the Spanish Civil War in 1936–37. Then came the outbreak of war in mid-1940, with the Italian Navy confronting an aggressively led Royal Navy Mediterranean Fleet. An early casualty was the *Bartolomeo Colleoni*, which joined her sister *Giovanni delle Bande Nere*. The two ships left Tripoli, Libya, on 17 July, bound for Leros in the Aegean, where British forces had been reported, but the squadron had already been sighted by a British reconnaissance aircraft, and early that morning, off Crete, they were caught by the Australian light cruiser HMAS *Sydney* and five destroyers. In the stern chase which ensued the Bartolomeo Colleoni was hit by a 6in shell in the engineroom and forced to stop, allowing the destroyers to close in and torpedo her. The *Giovanni delle Bande Nere* was lucky to escape.

Between September 1940 and March 1941 the *Alberico da Barbiano* was converted to a train-

The Bartolomeo Colleoni *during her engagement with HMAS* Sydney *and accompanying destroyers. Her bow has been blown off*

ing ship for the Naval Academy, suggesting that something was seriously wrong with her. In December 1941 she returned to the 4th Division as flagship for a special operation to re-supply the garrison in Libya. She and the *Alberto di Giussano* embarked petrol and munitions at Palermo and left on 12 December, but were ambushed by a force of British and Dutch destroyers off Cape Bon in the early hours of the following day. Both ships were torpedoed and blew up, a terrible setback for the Italian Navy.

The surviving member of the class, the *Giovanni delle Bande Nere*, had more success. After her escape from the guns of HMAS *Sydney* she continued to serve with the 4th Division on operations against the British. Her most successful feat was the laying of a minefield off Tripoli in June 1941, which later accounted for the cruiser HMS *Neptune* and the destroyer HMS *Kandahar*, and damaged the light cruisers *Aurora* and *Penelope*. She suffered damage in a storm on 23 March after an unsuccessful attack on the escort of a Malta convoy, and sailed from Messina on 1 April, bound for the main repair base at Taranto. Some 11 miles south of Stromboli she was spotted by HM Submarine *Urge*, which blew her apart with two torpedoes.

CONDOTTIERI CLASS FIRST GROUP

Alberto di Giussano laid down 29 March 1928, launched 21 December 1930, completed 5 February 1931, built by Ansaldo, Genoa

Alberico da Barbiano laid down 16 April 1928, launched 23 August 1930, completed 9 June 1931, built by Ansaldo, Genoa

Bartolomeo Colleoni laid down 21 June 1928; launched 21 December 1930; completed 10 February 1932; built by Ansaldo, Genoa

Giovanni delle Bande Nere laid down 31 October 1928, launched 27 April 1930, completed April 1931, built by Castellamare di Stabia

Displacement;	5110-5170 tonnes (standard), 6844 tonnes (full load)
Dimensions:	169.3m x 15.5m x 5.3m
Machinery:	2-shaft Belluzzo geared steam turbines, 95,000 shp; 6 Yarrow-Ansaldo boilers
Speed:	30kts (designed sea speed), 37kts (max, light)
Armament:	8-152.4mm 53 cal. Model 1926 guns (4 x 2), 6-100mm 47 cal. Model

	1927,(3 x 2), 8-37mm 54 cal. AA (4 x 2), replaced (1938-3) by 8-20mm,8-13.2mm MGs (4 x 2)m 4–533mm TT (2 x 2), 8 torpedoes embarked
Aircraft:	2 Cant 25 AR floatplanes (replaced by Ro 43s)
Range:	3800 nm @ 18kts
Fuel:	1150 tons oil
Complement:	507

A year after the order was placed for the Condottieri class, two improved versions were ordered, becoming known as the Condottieri Second Group. Although the intention was to provide better protection, the second group showed virtually no improvement apart from hull-strength and stability. The cumbersome aircraft hangar and catapult arrangements were also replaced, allowing the height of the bridgework to be reduced.

Main and secondary armament were unchanged, but the new Odero-Terni-Orlando Model 1929 152.4mm gun was substituted, with a roomier gun turret. Two single 40mm anti-aircraft guns were fitted, but these were replaced by four twin 20mm 65-cal. The catapult was sited abaft the after funnel, angled about 30 degrees to starboard, but no hangar was provided. In 1943 the survivor *Luigi Cadorna* had the catapult removed, and in 1944 her torpedo tubes were also removed. Both could be converted to lay 84 to 138 mines, depending on type.

The *Armando Diaz* was torpedoed by HM Submarine *Upright* on 25 February 1941 while escorting a convoy to Tripoli. Her sister survived and surrendered to the Mediterranean Fleet at Malta in September, and was then interned at Alexandria. Following the surrender of Taranto she returned, and operated under Allied control as a transport for Allied troops and equipment, and later for the repatriation of Italian PoWs from North Africa. Under the terms of the Peace Treaty signed on 10 February 1947 she was returned to the new *Marina Militare*. She was in poor shape and was used for training until stricken in May 1951.

CONDOTTIERI CLASS SECOND GROUP

Luigi Cadorna laid down 19 September 1930, launched 30 September 1931, completed 11 August 1933, built by CRDA at Trieste
Armando Diaz laid down 28 July 1930, launched 10 July 1932, completed 29 April 1933, built by Odero-Terni-Orlando at La Spezia

The basic design was further expanded for the 1930–31 programme, resulting in a Condottieri Third Group. The *Muzio Attendolo* and *Raimondo Montecuccoli* were some 2000 tons bigger than the Second Group, but had the same main armament. Protection was increased overall, making rather better cruisers out of them. During the Second World War both ships were active on convoy and support duties. In August 1942, while attacking the 'Pedestal' convoy, the *Muzio Attendolo* had her bows blown by a torpedo from HM Submarine *Unbroken*. Although she returned to Messina, and was then moved to Naples for repairs, she was destroyed in a daylight raid by USAAF bombers on 4 December that year, and her sister was badly damaged.

Post-war the *Raimondo Montecuccoli* was retained by the Italian Navy as a training ship, until stricken in 1964.

A Condottieri Fourth Group marked a further expansion, creating a much more capable design, although the main armament of eight 6in guns was repeated. The *Emanuele Filiberto Duca d'Aosta* survived the war but was ceded to Russia as the *Z.15*, then renamed *Stalingrad* (1949), and finally *Kerch*. She appears to have been sunk by mine off Odessa some time later, but the

The Fifth group of the Condottieri class was the best example of the type. The Giuseppe Garibaldi even survived until 1972 as a guided missile cruiser following a rebuild. (Line drawing 1/1250 scale.)

Soviet Navy never confirmed the loss. An alternative fate given is that she was repaired and served until 1957. The *Eugenio di Savoia* was ceded to Greece in 1951 and renamed *Elli*; she was sold for scrapping in 1964.

The Fifth Group was the final flourish of the Condottieri class, but the *Luigi di Savoia Duca degli Abruzzi* and *Giuseppe Garibaldi* bore little resemblance to the original design. Both were retained by the Italian Navy post-war, lasting until 1961 and 1972 respectively, and the *Giuseppe Garibaldi* underwent several rebuilds, culminating in a conversion to an anti-air warfare cruiser, armed with US-supplied missiles and electronics.

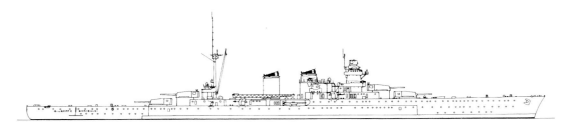

Conclusion

If remarks are confined to Groups One and Two, the Condottieri classes typify the Italian Navy's approach to warship design. Even today the Italian Navy is obsessed by achieving top speeds, irrespective of fuel consumption, and to that extent it continues to function as the spiritual descendant of Lord Fisher. Short range was understandable for ships intended to operate in the Mediterranean, but speed was not a substitute for armour protection, in either world war.

In one sense, the original Condottieri concept created by Generals Giuseppe Rota and Giuseppe Vian was more of a super-destroyer than a true cruiser. The light, virtually unarmoured hull was optimized for high smooth-water speed. The high quality of Italian ship-engineering guaranteed that they would be good steamers, but it is interesting to note that endurance at top speed went down to a theoretical maximum of 970 nm. Like other large destroyers, their light construction resulted in hull-damage when driven too hard in rough weather.

Policy papers found after the Italian surrender in September 1943 hint that Mussolini had no serious purpose, other than prestige, in building up his modern and superficially unbeatable Navy. This could be read another way; Mussolini may have believed that his armed forces and his threats of 'forests of Italian bayonets' would deter any foreign power from forcing an armed confrontation. That is explicable in the context of the moral weakness of the British and French governments of the 1930s. Neither nation had lifted a finger to stop or try to deter Italian military operations, and to Mussolini that was as good as a handshake.

The war-careers of the first six ships show that the Italian Navy did not skulk in its bases or run away at high speed, as suggested by British wartime propaganda. In fact all six saw considerable action, and had they not been faced by bold and aggressive Royal Navy senior officers they might have achieved strategic success. By dominating the Central Mediterranean, the Italian Navy would have had little trouble in subduing Malta and cutting the British supply-lines to the Middle East and India. The Italians also believed that their shore-based aircraft could dominate the Mediterranean, and if that was true, light anti-aircraft armaments embodied no risk. To an amateur strategist such as Il Duce, the task of securing shipping routes between southern Italian ports and bases and the North African colony of Libya did not require long range either. He is on record as saying that the Italian Navy had no need for aircraft carriers, as the whole of Italy was one big carrier!

The original Condottieri class were not suited to the job of ocean escort, being too lightly built. The efforts devoted to the substantial redesign of Groups Three to Five prove that the Italian Navy came to realise this as early as 1933, even before Groups One and Two could have been evaluated. The loss of five out of six ships in two years can be put down to the sort of random bad luck that occurs in wartime, but the misuse of the *Alberto di Giussano* and *Alberico da Barbiano* as fast transports at the end of 1941 reduced the odds even further. The somewhat facile pre-war predictions of naval warfare in the Mediterranean reflect poorly on a high command which had fought in the First World War. To be fair to the Supermarina, however, there was constant political interference, and many officers had been appointed for their connections with the Fascist movement rather than merit.

IJNS *Ryujo* Aircraft carrier

Imperial Japanese Navy 1929–42

To understand the paradoxes and contradictions inherent in the designs produced for the Imperial Japanese Navy (IJN) between the two World Wars, it is necessary to go back decades. After a brief civil war in the 1860s the rule of the Tokugawa clan Shoguns gave way to a more open system of selecting the country's officials using Western names and forms. With technical help from Europe the modern world was embraced fully. Germany helped to create a new army and Great Britain helped to build a modern Imperial Japanese Navy (IJN). With the growth of military efficiency, Japan joined the European nations in seeing the crumbling Chinese Empire as an easy path to enrichment. A short, sharp shock was administered to the Chinese in 1896, with the Imperial Navy's cruisers distinguishing themselves at the Battle of Yalu River and playing a vital role in transporting the Army to Manchuria. Then Russia, France and Germany intervened, forcing an end to the war and ordering Japan to withdraw from Manchuria. This was taken as a deadly insult, denying Japan what it had won by force of arms.

Then came the annihilation of the Imperial Russian Navy's Baltic Fleet at Tsushima in 1904 and the expulsion of the Russian Army from Manchuria. Japan had her place in the sun, and to the military hierarchy its armed forces seemed invincible. With alacrity Japan joined the Allies in 1914, and played a major role in capturing the German enclave in China at Kiaou Chou (Tsingtao), hunting for Admiral Graf von Spee's cruisers and escorting ANZAC troop convoys to Europe. Much more important was Japan's contribution to the Allies' war effort, in the form of shipbuilding. New shipyards were set up to handle large orders for merchant ships from the British, and a class of destroyers was built for the French.

The Army was still the dominant service and, in 1915 at the insistence of the high command, the Diet (Parliament) presented its infamous Fifteen Points to the Chinese Government; if implemented, these demands would have reduced China to a vassal-state of Japan. It had been assumed that the Western Powers would not intervene as they were preoccupied with the war, but the United States was not. Having already used its diplomatic clout to bring about an end to the Russo–Japanese war on terms that some in Japan thought were less than its due, the United States now intervened decisively, warning the Japanese Government to withdraw the obnoxious demands.

One cornerstone of Japanese foreign policy had rested on its alliance with Britain, agreed in 1902. The Anglo-Japanese relationship had been strategically fruitful for both sides. For Britain it created a counterweight to French and Russian ambitions in the Far East that was cheaper than British bases and British ships. For Japan it offered a ready source of technical assistance and an excellent model for how to run a navy, as well as an alliance with Russia's main rival.

The little Hosho *was Japan's first aircraft carrier, and the first in the world to be designed as such.*

However, in the 1920s all this changed when, as a consequence of the entry of the United States on the side of the Allies in the First World War, President Woodrow Wilson's schemes for disarmament took real shape in the form of the Washington Disarmament Conference of 1921. Under the resulting Naval Treaty, Britain and the United States were granted parity; Japan was not. Furthermore, under pressure from the United States, the British declined to renew their alliance with Japan. The grievances were piling up, and the combination of contempt for 'white' troops (the Japanese had only met badly-led Russian conscripts) and the need to expand was driving Japan down the road of aggression.

The start of serious interest in naval aviation by the IJN went back to 1912, when British experiments began to show promise. The Japanese were even more impressed by the rapid progress during the First World War, particularly the successful operations in 1918. As a result they ordered a small aircraft carrier in 1918, to be named *Hosho*. The ship was ordered in December 1919, becoming the first carrier in the world designed as such from the start, and she was completed only three years later. On a small displacement of 7470 tons she was a modest start, and could embark only 21 aircraft, of which 15 were to be operational with 6 spares. Speed was also unexceptional (25kts maximum) and her armament comprised four 5.5in low-angle guns and two single 3in for air defence. A sister, to be called *Shokaku*, was ordered in 1922, but was cancelled as a result of the Washington Naval Treaty.

As a result of the success of the *Hosho* the Navy Department decided to go for size to produce two carriers capable of operating with the battlefleet. Four large battlecruisers had been ordered, but they fell outside the 35,000 tons limitation imposed by the Treaty. Two were selected for conversion to carriers, the *Akagi* and *Amagi*, and funding was approved for the 1924 Programme. Work started on the *Amagi* at Yokosuka Dockyard in the summer of 1923, but before much had been done, disaster struck. On 1 September 1923 a huge earthquake devastated Tokyo and a wide area around. The hull frames of the 41,200-ton battlecruiser's hull were so badly strained and distorted that the conversion was stopped and the hull was broken up on the slipway. The *Akagi* was not affected as she was under construction at Kure, and the Navy Department decided to substitute the hull of the 39,930-ton battleship *Kaga*, also under sentence of death by the terms of the Treaty for the abandoned *Amagi*.

The two ships were not completed as identical carriers. The *Akagi* had no fewer than three flying-off decks, allowing aircraft to take off from the hangar without the longer procedure of using the two lifts to the flight deck. No bridge was provided, but this proved impracticable, and shortly after completion a small bridge was added on the starboard side of the upper flying-off deck. She was also armed with two twin 7.9in 50-cal gun turrets and six single guns in casemates aft. But to Western eyes the most bizarre aspect of the ship was the arrangement of funnels. The foremost funnel pointed outwards and downwards, combining the four uptakes

The Kaga, *as originally built, had large trunked funnels running astern along the hull. The smoke frequently drifted over the carrier deck, hampering operations.*

from the boilers, whereas the second funnel was upright and projected slightly above the flight deck. Nineteen boilers, eight coal-fired and 11 mixed-firing (i.e. coal or oil) developed over 130,000 shaft horsepower, good for 31kts.

Work started on the *Kaga* at the Kawasaki yard in Kure in 1924, but shortly afterwards the hull was towed to Yokosuka. Although the plans for the conversion drew heavily on the *Akagi* design, as a battleship the *Kaga*'s hull was about 67ft shorter and beamier. A different system of exhausting smoke was adopted, with two long horizontal trunked funnels, one on the port side and the other on the starboard side. Near the after end of the ship the two trunked uptakes were angled outwards. Both carriers had a capacity for 60 aircraft. Profiting by experience with the *Akagi*, a vestigial bridge (in fact no more than a conning position) was sited on the starboard side well forward. The main armament was identical, with two twin turrets on the centre flying-off deck and six in casemates aft. Because she was 4kts slower than her half-sister and shorter she was less satisfactory.

Both carriers were given massive rebuilds in the 1930s, and some of the major features were radically changed. They were given island superstructures, that of the *Akagi* being on the port side. In theory this would simplify the marshalling of aircraft after a strike was launched, with the two carriers operating side by side, rather than in line ahead. Known in Western navies as the 'D', the manoeuvring of the aircraft as they formed up for a strike, so the *Akagi*'s aircraft could form a port 'D' without the risk of getting mixed up with the other group. It was fine in theory, but the Royal Navy had experimented with a port-side island in 1918, and had discarded the idea very quickly. Although the position should not matter, experience showed that pilots tended to pull left on landing, and collide with the 'island'. Experience was to show that the Akagi's island likewise caused an increase in deck crashes.

Under the terms of the Washington Treaty the IJN was allowed to build a total of 80,000 tons of carriers, but most of the total had been used on converting the *Kaga* and *Akagi*. There was a loophole: carriers displacing less than 10,000 tons were exempt from the Treaty regulations, so the Naval Staff came up with the idea of building a small carrier, to be named *Ryujo*. The success of the two capital ship conversions marked the beginning of a uniquely Japanese approach to carrier design. The purpose of the carrier was seen as rapid approach to the battle zone, a surprise strike and an equally fast getaway. Hence the emphasis on high speed and the lack of importance attached to protection.

The draft design incorporated a single hangar, to keep within a limit of 8000 tons. But in the meantime experience with the first carrier, the *Hosho*, and the converted *Kaga* and *Akagi* showed that an aircraft carrier could only be effective if she could embark a minimum number of aircraft. As a result of these findings, the designers were told to incorporate a second hangar,

Superficially similar to the larger Japanese aircraft carriers, the Ryujo *in fact was a good example of how not to build an aircraft carrier.*

pushing the displacement up to 10,000 tons. To students of the IJN between the world wars it will come as no surprise to learn that this increase was not declared to the body supervising enforcement of the Treaty. This enabled uncritical analysts of the Japanese design-capability to provide yet another torrent of ill-informed praise, and gave them an opportunity to denigrate Western equivalents.

The *Ryujo* was completed in May 1933, and immediately problems appeared. She was unstable because of excess topweight, and hull-construction was too light to avoid distortions. Then came a brutal awakening for the designers; the torpedo boat *Tomodzuru* capsized in March 1934 in a gale while exercising off Sasebo. A thorough investigation into stability of all ship-types showed that topweight had been allowed to grow without any checks. As a result the *Ryujo* was returned to the dockyard for major modifications in August 1934.

The most important alteration was the fitting of larger bulges and a heavy ballast keel. At the same time two sets of twin 5in gun mountings aft were removed. She returned to service but more defects appeared. The hull was also strengthened, increasing displacement and reducing speed. These improvements went some way to remedy the ship's basic faults, but another weakness soon came to light. In 1935 the 4th Fleet suffered severe damage in a typhoon. In the storm *Ryujo*'s low forecastle caused her to ship massive amounts of water, endangering the ship's stability and causing potential damage to her electrical sub-systems.

Once again the carrier needed major modifications, and in May 1936 she returned to dockyard hands. The small bridge, lowered when launching and recovering aircraft, was reshaped to lower wind-resistance. The original close-range armament of 12.7mm heavy machine guns was replaced by twin 25mm. Eventually the problem of the water coming into the ship over the bow was rectified in 1940. Another deck was built on the original forecastle. As a result of all these defects the *Ryujo*'s reputation with the Fleet was poor. Operating aircraft from such a short flight deck was much slower than in other carriers. From a Japanese point of view the only positive outcome was to teach the designers how not to build an aircraft carrier, and much useful data was applied to the design of the much larger Hiryu class and the Shokaku class, widely acknowledged to be the most successful Japanese carriers.

When war broke out in December 1941 she was serving as the flagship of Carrier Division 4 and based at Palau, but was largely restricted to support operations, particularly during the invasion of the Philippines. Her air group at this period totalled 22 A5M4 'Kate' bombers and

THE WORLD'S WORST WARSHIPS

12 B5N2 'Claude' fighters, and they carried out the first strike against Legazpi on 13 December. In February 1942 her air group attacked the American-British-Dutch-Australian (ABDA) force east of Singapore, but scored no successes. She escorted the Java Invasion Force, but once again a strike against the ABDA force was unsuccessful. On 1 March 1942 her aircraft sank the old destroyer *Pope*. In April she and four heavy cruisers attacked Allied merchant ships in the Bay of Bengal. It was her finest moment, for her 12 'Kates' sank ten ships totalling 53,750 tons, although it was achieved against virtually no opposition. Her only 'real' front-line service was in the Eastern Solomons as part of the Diversionary Group, but it was also her last. At 1606 on 24 August 1942 she was sighted during the Battle of the Eastern Solomons. At 1620 she was attacked by aircraft from the USS *Saratoga*, and was set on fire by four 1000lb bomb-hits and mortally wounded by a torpedo. She was abandoned by her crew and sank about six hours later. Her aircraft were airborne at the time of the strike, and the survivors were diverted to a land base 400 miles away.

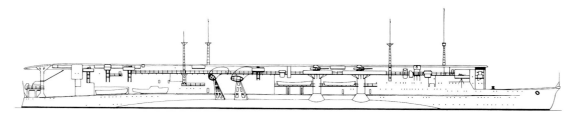

The Ryujo was too small for the kind of role the Imperial Japanese Navy envisaged for it. (Line drawing 1/1250 scale.)

IJNS Ryujo

Laid down 26 November 1929, launched 2 April 1931, completed 9 May 1933, built by Yokohama Dock Company and Yokosuka Dockyard

Displacement:	10,600 tons (standard)
Dimensions:	548ft 7in (pp), 590ft 4in (oa) x 66ft 8in x 18ft 3in flight deck: 513ft 6in x 75ft 6in
Machinery:	2-shaft geared steam turbines, 65,000 shp; six Kanpon boilers
Speed:	29kts
Armament:	8-5in 40-cal guns (4x2), 24-25mm AA guns 12 x 2)
Aircraft:	48 (36 operational, 12 spare)
Protection:	splinterproof only
Oil fuel:	2493 tons
Range:	10,000 nm @ 14kts
Complement:	924

Conclusion

The *Ryujo* was an inevitable victim of the IJN's obsession with trying to get a quart into a pint pot. The planned displacement of 8000 tons was ludicrously inadequate for a fleet carrier, and adding another hangar merely compounded the problem. The US Office of Naval Intelligence (ONI) personnel who cross-examined Japanese constructors in 1946 were only interested in technical assessments; there is no evidence of any interest in costs. Establishing comparable figures for costs of enemy warships is always very difficult, but we can say that the *Ryujo* must have cost far more than she was worth. If only in terms of time spent in dock undergoing rectification of design errors, she must be rated as a bad bargain.

There was one very serious flaw in all the carriers discussed above, and the later carriers as well: the precautions against aviation fuel fires. The author once told a largely American audience at a history conference at Annapolis: 'If the Japanese had designed your carriers you would have lost the Battle of Midway.' A superficially flippant comment, but it was based on the fact that the *Akagi*, *Hiryu*, *Kaga* and *Soryu* were first disabled and then destroyed by massive fires and explosions at Midway. In contrast, the USS *Yorktown* suffered severe bomb damage and fires, but survived long enough to play a vital role in the destruction of Admiral Nagumo's carrier force. The secret of US carrier design (and the Royal Navy's designs) was the elaborate care taken to protect and isolate the fuel tanks deep inside the ship. Critics may challenge this view by saying that the four Japanese carriers were destroyed by sheer weight of numbers. This does not bear close analysis because the US Navy's three carrier air groups suffered heavy losses trying to penetrate the Japanese combat air patrol before the battle suddenly started to go their way. Later other Japanese carriers suffered catastrophic aviation fuel fires, proving that the Japanese learned very little from their failure to capture Midway and destroy the Pacific Fleet's carriers.

With the benefit of hindsight the *Ryujo* might have been successful as an escort carrier, but in the 1930s no navy had refined such a concept. In any case the IJN would not have spent so much money on commerce-protection, a role which was despised as unworthy of a warrior nation. The Royal Navy was the only organisation to look seriously at conversions of liners to 'trade carriers'; the US Navy took the subject on board in 1940, resulting in large numbers of escort carriers (CVEs). Japan was never going to win a long war against the United States, but the disdain of the Navy for any measures to husband merchant shipping until 1943 was a major factor in the nation's downfall, even allowing for the nuclear attacks on Hiroshima and Nagasaki. The large Japanese mercantile fleet was a national asset, but it was treated as an expendable one, recalling the mistakes made by the British in 1915–16.

The sorry tale of the *Ryujo* reflects a constant headache for designers. It is all very well for the Treasury or Finance Ministry to try to drive down costs of warships by tempting the politicians to choose the cheapest option. But there is always a baseline below which a cheap, small design is not cost-effective. Warships are intended to fight, and a serious degradation of battleworthiness merely creates ships which cannot be risked in the front line. This does not mean that the only effective ships are the biggest, but the role of the ship and its ability to perform specific tasks with maximum efficiency must take precedence.

We also come back to the *raison d'être* of the aircraft carrier. Its purpose is to enable suitable aircraft to be operated far from land bases, so the starting point for designers must be the air group. All other measures, guns, armour and fire precautions are included only to ensure the effectiveness of the air group. Furthermore, all carrier experience to date proves that small carriers are less effective then big ones. A big hull can absorb more punishment if well-designed, and the safety of pilots when taking off and landing is easier to achieve. Big carriers can also operate longer because the greater internal volume permits more fuel and ordnance to be stowed.

Mogami class Cruisers

Imperial Japanese Navy 1930–1945

The near-annihilation of the Russian Baltic Fleet at the Battle of Tsushima was, as was discussed in the chapter on the Borodinos (see pages 53-5), a trial between British and French naval technology, much to the credit of the former. But there was a price to pay, and the Japanese became convinced that they were superior to any 'white' nation—after all, a nation which had been feudal less than 40 years earlier had destroyed Russian naval and land power. The Japanese believed the experience of this victory proved that any difference in the size of the respective fleets would be offset by superior individual fighting power and better training.

Although the IJN was steadily becoming self-sufficient in the production of guns and armour, it still relied on British industry for the heaviest guns and armour. Then came the outbreak of the First World War in 1914, and for the first time the British became conscious of the need to tighten technical security. During the war the Royal Navy's warships increased greatly in fighting power, and by the end in 1918 plans for much larger capital ships existed, but the Japanese were not given access to any of these major developments.

The Washington Naval Disarmament Conference of 1921 and the resulting treaty of 1922 was a serious blow to the Japanese, who complained bitterly of the 'Rolls-Royce:Rolls-Royce:Ford' ratios of capital ships permitted—parity for the US and Royal Navies, but second-rank status for the IJN. Worse, it imposed straitjackets on the tonnages allocated for new major warships: 35,000 tons and 16in guns for battleships and 10,000 tons and 8in guns for cruisers.

The Japanese had reason to feel threatened. The United States, had detached Britain from its Far Eastern ally. The United States had relegated Japanese, a member of the same alliance that had won the First World War, to second-rank naval status. The increasingly militarisic governments believed that conflict with the US Navy and Army was inevitable. The future promised more tension and an arms race. But this race would be started without the technical guidance of the Royal Navy and the wartime experience mentioned. The IJN's constructors would have a largely clean sheet. To make matters worse, the constructors were ordered to approach each design without reference to previous experience. This was intended to 'free the mind' to allow each new class to be superior to anything afloat—fine in theory but running the risk of allowing size and cost to run out of control very quickly.

On 5 February 1922, the day before the signing of the Washington Treaty the Japanese Government stopped work on all capital ships, but accelerated work on any ships not covered by the Treaty, in order to keep the shipyards employed. These included two 7500-ton 'large scout cruisers', but at the earliest possible moment a new Naval Armament Limitations Programme was drawn up, adding two more 7500-ton and four 10,000-ton Class A heavy cruisers.

The Mogami *during her trials in 1935. The lightly constructed hull suffered distortions and loose plating.*

The Staff Requirement for the smaller cruisers, to be known as the Furutaka class, were demanding: a normal displacement of 7500 tons, an armament of six single 8in guns and six 24in torpedo-tubes and a speed of 35kts. During construction of the *Furutaka* and her sister *Kako* the Navy General Staff nagged the Basic Design Section to alter the main armament to three twin 8in gun mountings and to make other changes. They were, however, too advanced for such a major change, so the alterations were made to the second pair, *Aoba* and *Kinugasa*. Unfortunately there was a small drawback: the alterations would increase the two-thirds trials displacement from 8586 tons to 8910 tons (actually 9502–9820 tons). The 3in side armour was only proof against 6in gunfire at 12,000–15,000 yards, but they all reached their designed speed at a displacement of 8600 tons.

Other navies were not told of the difficulties in achieving the final design, and took the original figures as genuine. As late as the 1930s the Royal Navy's Director of Naval Intelligence (DNI) was berating the DNC for building 8in gun cruisers so inferior to the Kinugasa class. To which the DNC replied that the Japanese could not achieve the qualities on such a low displacement unless they were building their ships of cardboard, or more likely, the DNI's figures were wrong.

The IJN found it increasingly hard to work within the 10,000-ton limit, and the staff succumbed to the temptation of quoting publicly only the intended displacement, and keeping quiet, i.e. lying, about the actual figure. After the London Naval Treaty of 1930, which set tonnage totals for heavy (8in guns) and light cruisers (6.1in guns), the IJN was allowed a total of 108,400 tons for heavy cruisers and 100,450 tons for light cruisers. The completion of the last of the current class of heavy cruisers would use up the IJN's tonnage allowance, whereas by deleting over-age vessels there was a further 50,955 tons available for light cruisers. To use up this total the Naval General Staff planned to build four 8500-ton ships, followed by two 8450-ton ships.

Orders for the four 'medium cruisers' were placed in 1931 (the first pair) and 1933 (the second pair), to be armed with 15-15.5cm guns in triple turrets, and a speed of 37kts. The *Mogami* was laid down at Kure Dockyard in October 1931, followed by the *Mikuma* at the Mitsubishi yard in Nagasaki two months later. The *Suzuya* was laid down at Yokosuka Dockyard in December 1933, followed by the *Kumano* at the Kawasaki yard in Kobe in April 1934. It was hoped that the *Suzuya* would be commissioned in January 1936 but trials were interrupted in November 1935 after the alarming analysis of the so-called Fourth Fleet Incident. In

The view forward from the superstructure of IJNS Mikuma *showing the original armament of three triple turrets with 15.5cm guns. These were removed and replaced with 20.3cm guns on all ships of the class in 1939-40.*

1934 the torpedo boat *Tomodzuru* had capsized in a gale at Sasebo, drawing attention to the deficiencies in topweight brought about by 'Beat the Treaty' measures.

In the light of the findings of the Board of Enquiry, all ships ordered in the First Replenishment Programme were scrutinised closely for any lack of stability. The *Mogami* and *Mikuma* had their bridgework reduced in height and weight, and the seaplane hangar and various deck structures were deleted. In March 1935 the *Mogami* started her trials, without catapults, main armament directors and part of her anti-aircraft armament, but still displaced 12,669 tons. During her full-power trials, frames and side stringers aft were distorted, the hull-plating was loosened and fuel tanks were ruptured. The movements of the lightly constructed hull also distorted the roller paths under the turrets. Following emergency repairs at Kure, the *Mogami* and her sister *Mikuma* joined the fleet. Then in September 1935, during a Fourth Fleet exercise in which the big destroyer *Hatsuyuki* had her bows wrecked as far back as the bridge in bad weather, both cruisers suffered further hull distortions. Electrically welded seams split and the 6in turrets could not be trained because of hull-distortion. On their return to Kure both were placed in 'second reserve' on 15 November 1935. The trials of the third ship, *Suzuya*, were stopped by order of the Extraordinary Ship Performance Board, and she too went into reserve, while work on the Kumano was suspended.

Following the findings of the Board, measures to increase longitudinal strength in all ships which had been electrically welded were taken, and between 1936 and 1938 the first three

Mogamis were reconstructed. Welded shell-plating was replaced by riveted plating for 80 per cent of the length, although the ends were of welded mild steel. Additional Ducol steel plating was added amidships to the bottom plating on either side of the keel, and on the AA gun deck and sides, to reduce the risk of damage to the gun turrets. Expanded bulges were fitted to improve transverse stability, and minor changes were made, such as reducing the height of the mainmast. Unfortunately displacement now averaged 13,000 tons in trials condition, 30 per cent heavier than the Treaty limit.

Japan was becoming increasingly unhappy with the naval treaties, and on 29 December 1934 denounced the Washington Treaty. This was followed by the decision of 15 January 1936 not to sign the Second London Naval Treaty, and as a result Japan entered into a 'Non-treaty Period' on 1 January 1937. This allowed the IJN to up-gun the *Mogami* class and their successors, the two *Tone* class, with twin 20.3cm guns instead of triple 15.5cm.

The major technical problem was the need for larger roller paths to support the larger turrets. The new 20.3cm 50-cal Type 3 No.2 guns had a maximum range of 29,400m and fired a 125.85kg shell capable of penetrating 150mm armour at 15,000m. The turrets were an improved version of the 'E' type mounted in the *Maya*, modified to fit on a bigger roller path.

When she finally completed her repairs and upgrading, the *Mogami* was assigned to *Sentai* (squadron) 7 and was joined by her sisters as they emerged from the dockyards. Their first operational deployment was to Hainan Island in January 1941, to deter the French squadron based at Saigon, French Indochina, from participating in the skirmishes concerning a border dispute with Thailand.

When war broke out with the United States and its allies the four ships embarked on a war that would get harder as time passed, and saw action in all the major battles. The first was the conquest of the East Indies after the fall of Singapore; in the battle of the Java Sea *Sentai* 7 sank the USS *Houston* and HMAS *Perth* with gunfire and torpedoes. The *Mikuma* was the first of the four to be sunk, after being accidentally rammed by the *Mogami* during the Battle of Midway, then being attacked by a US Navy carrier aircraft on 7 June 1942 and turned into a blazing wreck. The *Mogami* was also badly damaged, but managed to limp back to Truk. The *Kumano* and *Suzuya* fought in the Solomons campaign, and were then refitted and given extra anti-aircraft weapons.

The *Mogami* emerged from major repairs and upgrading of her close-range armament at the end of April 1943. The three remaining ships fought in the Battle of the Philippine Sea, but their charmed lives were drawing to a close. The US invasion of the Philippines in October 1944 brought on the great battle of Leyte Gulf (actually a series of separate actions). The *Mogami* was sunk by gunfire and torpedoes during the disastrous night Battle of Surigao Strait on 25 October, and was found the next morning by carrier aircraft. The *Suzuya* was sunk by carrier aircraft on 25 October while trying to attack the invasion transports. The *Kumano* was sunk by carrier aircraft next day in the Battle of the Sibuyan Sea.

Mogami class

Mogami laid down 1 August 1932, launched 14 March 1934, commissioned 28 July 1935
Mikuma laid down 24 December 1931, launched 31 May 1934, commissioned 29 August 1935
Suzuya laid down 11 December 1933, launched 20 November 1934, commissioned January 1936 (put into reserve pending reconstruction; officially commissioned with *Kumano* on 31 October 1937
Kumano laid down 5 April 1934, launched 15 October 1936, commissioned 31 October 1937

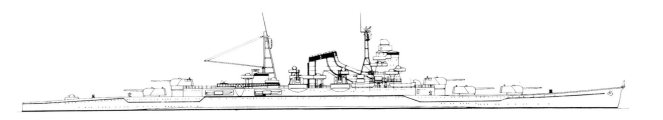

The Kumano *in 1944 after additional anti-aircraft armament had been installed. (Line drawing 1/1250 scale.)*

Displacement:	12,962 tons (trials, 1935)
Dimensions:	189m (oa) x 18m x 6.1m (mean)
Machinery:	4-shaft Kampon geared steam turbines, 10 Kampon boilers; 152,000shp
Speed:	37kts (designed)
Armour:	25-140mm belt; 35-40mm deck
Armament:	15–15.5cm 60 cal. guns (5 x 3), 16–12.7cm 40 ca. AA guns (8 x 2), 16–25mm AA guns (8 x 2); 8–13mm MGs (4 x 2), 12–61cm torpedo-tubes (4 x 3)
Aircraft:	Deck stowage for 3 Kure No.2 model floatplanes
Complement:	70 officers, 780 ratings

(AFTER FIRST RECONSTRUCTION)

Displacement:	13,887 tons (2/3 for trials)
Dimensions:	189m (oa) x 18m x 7.6m (mean)
Machinery:	unaltered
Speed:	35kts (as completed)
Armour:	unaltered
Armament:	10–20cm 50 cal guns (5 x 2), 10–12cm AA guns (8 x 2); 10–25mm AA (8 x 2), 10–13mm MGs (8 x 2), 12–61cm torpedo tubes (4 x 3) (up to 24 Type 90 Mk 1 torpedoes carried)
Aircraft:	4 Kure Type No.2 floatplanes
Complement:	58 officers, 893 ratings

Conclusion

The problems encountered in the Mogami class demonstrate the limits imposed by the laws of physics and hydrodynamics on naval architects. When trying to 'cram a quart into a pint pot' there is a limit on what can be achieved by clever design. The Naval Staff's insistence that each new class must be superior to all existing opponents placed an enormous burden on the IJN's constructors, and we must judge their successes and failures against that background.

The US Navy's ONI personnel who interrogated Japanese constructors in 1946 were amazed at the innovations in IJN designs. They were, however, equally surprised at the number of simple mistakes made, and attributed these to the deliberate lack of reference to previous experience. One such failure was the practice, mainly in cruisers, of designing the 'run aft' of the hullform to give maximum speed, without allowing for the downward 'tuck' of the stern as the ship worked up to full speed. This meant that the ship did not develop her theoretical maximum power. Another puzzle for ONI teams was the deliberate discontinuity of the main girder—that 'wavy' deck line so obvious in photographs. US and British design practice avoided such discontinuity, and regarded it as incompatible with ultra-light construction.

Until 1942 the IJN frightened its opponents, but thereafter its warships' weaknesses become better understood, and it is significant that the US Navy never adopted any of its design procedures, during or after the Second World War. Critics might ask why, and the answer is that the US Navy produced the most powerful *and* the most reliable warships in all categories. Its ships were also easier to build, sticking to a tradition established by the British in their long war against the French Revolution and Napoleon's First Empire. Only an inferior foe feels the need to find an 'equaliser', and it has not yet staved off final defeat for any navy.

Yamato class Super battleships

Imperial Japanese Navy 1937–1944

The Imperial Japanese Navy felt cheated by the Americans and British at the Washington Conference in 1921, when it was denied parity in capital ships and heavy cruisers. When Great Britain, under pressure from the United States, refused to renew the Anglo-Japanese Treaty, this confirmed the suspicions of the military high command that Japan's security and prosperity must not be based on any amicable relations with the Western nations. Unfortunately, any major improvement in Japanese economic and industrial prosperity could only come in some form from outside. The Land of the Rising Sun was desperately short of agricultural land to feed its growing population, and apart from poor-quality coal, lacked raw materials to expand industry.

The end of hostilities in 1918 hit Japan hard. Its export markets were now moving to protectionism and in any case the post-war slump resulted in severe hardship. The shipyards became desperately short of work, and the huge capital ships under construction to challenge American and British seapower were becoming unaffordable. The Washington Treaty allowed the conversion of two large hulls to aircraft carriers, but the 1923 Tokyo Earthquake inflicted severe and lasting damage to the local shipyards.

Although near-miracles were achieved in rebuilding the existing capital ships, it was clear to the Naval Staff that the ships could not undergo improvements in fighting power indefinitely. The fleet could not, therefore, hope to have a decisive margin over modern American or British battleships in a fleet action. Out of this was born the germ of an idea to build battleships outside the limits imposed by the Washington Treaty and its successors. If there were no limits on gun-calibre and standard displacement, a truly decisive margin of offence and defence could be guaranteed.

Design work started in 1932 and the first complete design was presented in March 1935, but it went through another twenty-two stages before a final design was approved in 1937. The date was important because the London Navy Treaty of 1936 had renewed the tonnage and gun-calibre limits set at Washington fourteen years earlier. With a fine mixture of naivety and duplicity, the Japanese Naval Staff convinced itself that if Japan refused to accept the London Treaty, it would not be guilty of cheating if it built two much larger ships which were not completed until 1940. It was also reasoned that the US Navy would be unable to build comparable ships because they would have too much beam to transit the Panama Canal locks, or that they might even be able to frighten the Americans into giving up the struggle and conceding Japanese hegemony in the Western Pacific. Obviously the strategy would only work if the signatories to

The IJNS Yamato *was an enormous battleship, designed for a refight of Tsushima against the USN, in an era when the aircraft had added a third dimension to the conditions of a 'decisive battle' like Tsushima.*

the Treaty, particularly the Americans, had no idea of how big the ships would be. The other navies were, in fact, led to believe that the ships would displace no more than 40,000 tons and would be armed with 41cm guns.

It was hoped to propel the ships with a combination of steam turbines developing a total of 75,000 shp, and two-cycle double-acting diesels developing 60,000 bhp. The installation was 2500 tons heavier than an all-turbine installation, but offered much lower fuel-consumption. This design was ready in July 1936, but two months later the prototype diesel installation in the big submarine depot ship *Taigei* developed a serious design-fault. As the diesels in the new battleships were to be covered by 8in armour they would not be able to be repaired without a major refit, and as a result the plant had to be altered to an all-turbine installation. This final plan was drawn up in March 1937.

The first two ships were ordered under the 1937 Third Reinforcement Programme, while two more, one to be named *Shinano* and the other known as *Hull No. 111*, were ordered under the 1939 Fourth Reinforcement Programme. A fifth ship was planned as *Hull No. 797* but never laid down. The design included the most elaborate protection systems, including inclining the belt at 20 degrees and subdividing the hull into 1147 watertight compartments, of which 1065 were below the armoured deck. All turbines, boilers and auxiliary machinery were in individual compartments. The hull was unusually beamy, with a length:beam ratio of 27:4. The decks were calculated to be proof against 1100lb bombs dropped from 11,000ft, while the massive side armour was intended to keep out 18in shells to a distance of 16 nautical miles.

Building such huge hulls created unusual problems. The overriding need for secrecy was met by surrounding the building berths at Kure and Nagasaki with huge curtains of netting. The quantities of hemp required to manufacture the immense screens dislocated the fishing industry, which could not replace its nets. At Kure a small hill overlooking the building berth was levelled to deny a vantage point to friendly or unfriendly snoopers. Rolling and machining of the armour plates proved troublesome; cracks appeared at the corners, so smaller plates had to be substituted.

PARTICULARS OF THE 46CM GUN

Weight: 165,000kg
Weight of shell: 1460kg
Range: 42,030 yards @ 45 degrees
Rate of fire: 1 round every 90 seconds
Weight of turret: 2470 tonnes

Special arrangements had to be made for transporting the guns and turret machinery, and for lifting them into the ships. A merchant ship was converted for the purpose. The blast from the

The huge guns of the Yamato class were intended to outrange any foe and penetrate any armour.

46cm guns proved lethal. Not only did it inflict structural damage but also injury on personnel standing too close. Firing trials showed that blast ripped clothing off and tore flesh, as well as causing permanent hearing loss. As a result the 12.7cm AA mountings had to be fitted with blast-proof shields.

Despite the impressive success of the carrier air strike on Pearl Harbor on 7 December 1941, events did not conform to the pre-war strategy envisaged by the Naval Staff. There was no fleet action for the *Yamato* and *Musashi* to use their awesome firepower (there were not enough US Navy battleships left in the Pacific), and instead the US aircraft carriers staged a series of raids from Pearl Harbor. These did not produce decisive results, but they distracted the Naval Staff from its objective of sweeping the Pacific clear of US forces. The Battle of the Coral Sea in May 1942, for example, frustrated a Japanese thrust to attack Australia. The Commander-in Chief, Admiral Isoroku Yamamoto, decided to bait an elaborate trap to catch all the US Pacific Fleet carriers, to rid himself of this nuisance. As history records, the Battle of Midway in June 1942 turned the tables by eliminating four of Yamamoto's carriers, for the loss of only one American, the USS *Yorktown*. Yamamoto and the Central Fleet were unable to save the carriers, and the presence of the flagship Yamato made no difference to the outcome.

The decisive fleet battle continued to elude Yamamoto, as the air threat continued to escalate. The wing triple 15.5cm turrets (ex-Mogami class), were removed from both ships in the

The Shinano, *converted from a battleship hull to an aircraft carrier, drawn by S. Fukui. The ship had a short life, being torpedoed by a USN submarine while on her trials.*

autumn of 1943, and replaced by 12 triple 25mm mountings. It had been hoped to add six twin 12.7cm mountings, but this was only done to *Yamato* while she was under repair at Kure after being torpedoed by the USS *Skate*. At the same time twelve more triple 25mm and 26 single 25mm were added.

The extra 12.7cm gun mountings were not available for the *Musashi*, so she received 12 triple 25mm mountings and 25 singles, while under repair at Kure after being torpedoed by the USS *Tunny* on 29 March 1944. She and her sister took part in the disastrous Battle of Leyte Gulf. On 24 October 1944, while steaming through the Sibuyan Sea, she was attacked by aircraft from the carriers *Cabot, Enterprise, Essex, Franklin* and *Intrepid*. Under the remorseless hail of bombs and torpedoes she succumbed to progressive flooding and sank at 1935.

The *Yamato* suffered bomb damage and took on a lot of water during the Battle of Leyte Gulf, but escaped with the remnants of the Japanese fleet. Thereafter she remained in home waters, virtually immobilized for lack of fuel. When US forces landed on Okinawa she was earmarked for the ultimate kamikaze attack, and was ordered to beach herself on Okinawa and use her guns as a coast defence battery. Enough fuel was available for a one-way trip, but the pride of the Imperial Japanese Navy was never to complete her mission. On 7 April 1945 she was spotted by aircraft of Task Force 38 while still 138 miles west south west of Kagoshima. The carriers *Bataan, Belleau Wood, Bennington, Bunker Hill, Cabot, Essex, Hornet, San Jacinto* and *Wasp* sank her with an estimated 11 torpedoes and seven bombs. As she sank below the surface a massive explosion took place, probably caused by fused shells falling out of their racks.

The story of the third unit, the *Shinano*, was even more bizarre. She had been laid down under the Fourth Reinforcement Programme, but late in 1940 work slowed down. By June 1942, at the time of the Battle of Midway, it was clear to the Naval Staff that the need for new battleships had been overtaken by the requirement for as many aircraft carriers as possible. The hull of the *Shinano*, complete up to the main deck, was seen as suitable for conversion. But even at this late stage there were two schools of thought about the conversion; one faction wanted a full conversion, while another wanted a partial conversion to allow the ship to be used as a huge mobile support ship for existing carriers. Eventually a compromise was agreed, a 68,000-ton carrier carrying 24 aircraft for self-defence (plus seven for spares), and equipped with workshops, stowage for spares, fuel, ordnance and even food.

The ship was finally launched in October 1944 and was commissioned for trials. She was ordered to Matsuyama in the Inland Sea to complete fitting-out away from the devastating air attacks. At 03:17 on 29 November she was stalked by the USS *Archerfish* and suffered four torpedo-hits. The damage was not fatal, but her crew and the shipyard workers on board were not familiar with the layout. A number of watertight doors were left open, valves failed to function, and poorly welded seams gave way. Seven hours after being hit the mighty *Shinano* rolled over and sank.

The Yamato *as she looked in 1945, just before her ill-fated* kamikaze *run toward Okinawa. (Line drawing 1/2400 scale.)*

YAMATO CLASS

Yamato laid down 1937, launched 8 August 1940, completed 16 December 1941, built by Kure Dockyard

Musashi laid down 1937, launched 1 November 1940, completed 5 August 1942, built by Mitsubishi, Nagasaki

Shinano laid down 4 May 1940, launched 8 October 1944, completed for trials 19 November 1944, built by Yokosuka Dockyard

Displacement:	64,000 tons (standard), 67,123 tons (normal), 71,659 tons (full load)
Dimensions:	263m (oa), 244m (pp) x 36.9m x 10.4m (max)
Machinery:	4-shaft Kampon geared steam turbines; 12 Kampon boilers; 150,000 shp
Speed:	27kts
Armament:	9-46cm 45-cal guns (3 x 3) 12-15.5cm 55-ca. guns (4 x 3), 12-5in 40-ca AA guns (6x2), 24-25mm AA guns (8x3), 4-13mm MGs (4x1)
Aircraft:	7 floatplanes, 2 catapults
Armour:	16.1in belt; 11.8in bulkheads; 9.1in decks, 11.8in torpedo bulkheads, barbettes 21.5-22in; turrets 25.6-9.8in; CT 19.7-9.8in
Fuel:	6300 tons oil
Range:	7200 nm @ 16kts
Complement:	2800 (maximum)

CONCLUSION

The story of the *Yamato* and her sisters is an outstanding example of the isolation of the Japanese military caste from reality. The argument about building ships in secret and completing them just after the expiry of the London Treaty sounds more like the machinations of a firm of lawyers than the deliberations of admirals. As for the notion that the United States could be frightened into accepting priority with Japan, the Japanese were not the first or the last to underrate 'effete' democracies.

The enormous industrial effort required to build such massive ships was almost beyond the capacity of the Japanese economy, and limited the number which could be built. The money spent could easily have bought three or four carriers, but the Naval Staff was obsessed with the memory of Tsushima, and sought the decisive 'battle of annihilation'. This obsession co-existed with a very progressive approach to carrier warfare.

The problems arising out of the sheer size of the hull and the main armament have already been touched on, and it is clear that the pursuit of size had reached a point where it was inhibiting the designers. Such deep-draught ships required deep anchorages and special docks (in an age which knew nothing of super-tankers). This reduced the flexibility of deployment, one of the major advantages of warships.

Yet again, an attempt to build a 'super-ship' failed. The *Yamato* and *Musashi* achieved nothing more than the older and smaller battleships of the fleet did. The great 'white elephant' *Shinano* was final proof that the Naval Staff's pre-war planning was totally flawed. No navy got everything right before the Second World War, but the Imperial Japanese Navy got more wrong than most. The dedication of its sailors and aircrew could only postpone the inevitable, aided by some bad mistakes by its opponents.

BISMARCK CLASS BATTLESHIPS

GERMAN NAVY 1939–1944

In the light of the extraordinary admiration for the *Bismarck* and her sister *Tirpitz*, their inclusion in the ranks of the world's worst warships might seem eccentric or downright perverse. The real facts, however, point in the opposite direction to the widespread reverential attitude to any example of German technology, particularly that of the Hitler era.

In a sense, the Royal Navy was as much to blame as the media and Nazi propaganda. After all, she sank the pride of the Royal Navy, HMS *Hood*, and no allowance is made for the discrepancy of age and design philosophy. The Navy and the public would not have been human if they did not magnify the *Bismarck*'s fighting qualities, in much the same way that the US Navy assumed that the Japanese had some secret method of achieving design superiority. Neither hypothesis had much basis in fact, but to this day the myth that 'the Bismarck was not sunk' overrides any objective assessment.

In another sense the story goes back to 1919, when the interned High Seas Fleet scuttled itself ignominiously at Scapa Flow. The shame of Scapa Flow was deeply ingrained in the ethos of the postwar *Reichsmarine*'s officer corps, and this led logically to a dream of a powerful surface fleet capable of meeting the Royal Navy and defeating it in battle, an updated version of the pre-1914 dream of *Der Tag*, a trial of strength with the biggest navy in the world. But the mission for the new capital ships envisaged by the Navy's Commander-in-Chief Admiral Erich Raeder, was not to engage the main strength, but to operate as commerce raiders, disrupting the convoys of merchant ships to allow the U-boats to attack from a position of strength. The presence of a major unit would also frighten off the escorts and inhibit their ability to hunt the U-boats—at least, that was the theory.

Hitler was anxious to win the loyalty of the armed forces, and shortly after coming to power in 1933 funding was provided for the building of battleships F (*Ersatz Hannover*) and G (*Ersatz Schleswig-Holstein*), to be designed by the Navy's chief naval architect, Burkhardt.

Design work started in 1933 and was completed in 1936, followed almost immediately by the laying down of both ships. The design was impressive: 79 per cent of the flush-decked hull was welded, subdivided into 22 compartments, having a double bottom over 83 per cent of the keel. Partly because of the urgency and partly because of the 22-year gap since the last battleship had been designed, the Bayern of 1914 was used as the basis of the new design. This fact was to prove of great significance later, as we shall see, and cannot be ignored or underestimated. Even the main armament was the same as the Bayern: 4 twin 38cm gun turrets, although the new ships had an improved 47-cal gun. The main difference was the secondary armament of twelve 15cm guns, mounted in twin turrets at weather deck level.

The Bismarck *as seen from Prinz Eugen. Bismarck's flaws were perhaps to be expected in a navy whose experience of building large capital ships had undergone a twenty-year hiatus.*

The first ship, named *Bismarck* at her launch early in 1939, went to sea in April 1940. Her sister *Tirpitz* was launched in the spring of 1939 and commissioned early in 1941. The former then began her work-up in the Baltic, preparing for a raid against the North Atlantic convoys accompanied by the heavy cruiser *Prinz Eugen*. A total security blackout on Operation Rheinübung (Rhine Exercise) was imposed to conceal the breakout from the British, but it was compromised almost before it began. The British Naval Attaché in neutral Stockholm, Captain Denham, was alerted by a sympathetic Swedish naval officer (a rarity in largely pro-Nazi Sweden), who told him that two large warships had been sighted by the Swedish cruiser HSwMS *Gotland* (qv) heading for the Kattegat and eventually the open sea. They could only be the *Bismarck* and *Prinz Eugen*, and it was correctly assumed that they were heading for Norway. Immediately air reconnaissance missions were flown over Bergen and Oslo, and at 13:15 on 21 May a Spitfire of Coastal Command on photographic reconnaissance spotted the German ships refuelling in Bergenfjord.

The British were now aware of the Atlantic foray, and just before midnight the *Hood* and *Prince of Wales* left Scapa Flow, heading for Iceland, where they would refuel before taking up station to the south of Iceland. From there they could cover the Iceland-Greenland and Iceland-Faeroes Gaps. In addition the two heavy cruisers *Norfolk* and *Suffolk* patrolling the Denmark Strait were alerted.

That afternoon the *Prinz Eugen* had refuelled at Kalvanes, but the *Bismarck* did not, despite having burned over 1000 tons since leaving Gotenhafen. The reason for this apparently incompetent staffwork was that the refuelling hose had broken, spilling a large quantity of furnace oil. It was not a good start to Rheinübung, and left no margin for error.

The broad details of the engagement on 24 May we have already discussed on page 99 and will not be repeated. In the early hours the *Bismarck* and *Prinz Eugen* engaged and sank the British flagship HMS *Hood* and then turned their guns on her consort, HMS *Prince of Wales*. The second phase is generally assumed to have been very one-sided, one German account referring to *Das Cowardschiff*, implying that the *Prince of Wales* fled from the scene. In fact the *Prince of Wales* fought back and scored three hits on the *Bismarck* before her captain was ordered by a senior officer to break off the action. She stood up well to the German ships, absorbing seven hits from 38cm and 20.3cm shells. The most damage was from a 38cm hit on her compass platform, which killed or wounded everyone there; the shell did not explode but ricocheted off the binnacle and exited on the other side. When the *Prince of Wales* was docked for repairs, it turned out that none of the other hits had detonated properly, suggesting that the Kriegsmarine's base-fused heavy shells were not performing as designed. What the British did not know was that one of the *Prince of Wales*'s 14in hits had damaged a fuel bunker, contaminating the fuel and further reducing the *Bismarck*'s endurance.

When the *Prince of Wales* and her escorts disengaged, the cruisers *Norfolk* and *Suffolk* resumed their shadowing, with the battleship in support. In addition to being a radar echo, the German battleship was now leaking oil and had one oiler knocked out. Her only hope was to reach Brest in northwest France, 600 miles away, for she had little chance of making a rendezvous with an oiler. The following evening Admiral Lütjens made a diversionary lunge at his shadowers to allow the *Prinz Eugen* to head directly for Brest.

The Royal Navy was making an all-out effort to prevent the *Bismarck* from doubling back to Germany or reaching the comparative safety of Brest. Until heavy units could be concentrated the only hope of slowing her down was a torpedo attack. The first was a strike by nine slow Swordfish biplane bombers, and two Fulmar shadowers, launched from the flight deck of HMS *Victorious*.

Although the raid seemed suicidal, no Swordfish was shot down. A single torpedo hit, but it exploded close to the surface on the battleship's main armour belt, expending most of its force upwards rather than inwards.

Shortly after this lucky escape the *Bismarck* had another stroke of luck. The shadowing heavy cruiser HMS *Suffolk* lost radar contact, a fact which was known to the cryptographers of the B-dienst ashore, but not recognised by Lütjens' staff. This was a result of combination of poor technical intelligence about the range of Royal Navy radars and the mistake of assuming that the radar pulses detected had sufficient energy to return to the receiving antenna in HMS *Suffolk*.

During the ensuing three days, the German battleship was attacked twice by aircraft carriers. The second strike, launched by HMS *Ark Royal*, wrecked the *Bismarck*'s rudder with a torpedo, making her almost unmanoeuvrable. She beat off a night attack by destroyers, but the watching cruisers kept up a stream of reports, and early in the morning of 27 May she was delivered to the guns of the Home Fleet, the flagship HMS *King George V* and her consort HMS *Rodney*. The *Bismarck* was silenced within half an hour, but the waterlogged wreck did not sink until torpedoed by the heavy cruiser HMS *Dorsetshire*. She took with her 1,977 men, and just over 100 survivors were rescued.

BISMARCK CLASS
Bismarck laid down 01 July 1936, launched 14 February 1939, commissioned 24 August 1940

Tirpitz laid down 20 October 1936, launched 1 April 1939, commissioned 25 February 1941

The Bismarck's *handsome profile hid a multitude of obsolescent design features. (Line drawing 1/2400 scale.)*

Bismarck

Displacement:	45,950 tons (designed); 41,700 tons (standard); 50,300 tons (deep load)
Dimensions:	248m (oa), 241.5m (wl) x 36m x 10.6m (maximum)
Machinery:	3-shaft Blohm & Voss geared steam turbines; 12 Wagner boilers, 138,000shp (designed)
Speed:	29kts (designed), 19kts (cruising)
Armour:	12.5in belt, 4.8in deck (maximum)
Armament:	Armament: 8-38cm L/47 guns (4 x 2), 12-15cm L/55 guns (4 x 2), 16-10.5cm AA guns (8 x 2), 16-3.7cm L/83 AA guns (8 x 2), 12-2cm AA guns (12 x 1)
Aircraft:	2 Ar.196 floatplanes
Complement:	c.108 officers, 2,500 ratings

Tirpitz

Displacement:	42,900 tons (designed); 52,600 tons (maximum)
Dimensions:	As *Bismarck*
Machinery:	3-shaft Brown Boveri geared steam turbines, otherwise as *Bismarck*
Speed:	as *Bismarck*
Armament:	as *Bismarck*, but 16-58 2cm AA guns
Aircraft:	6 Ar.196 floatplanes
Complement:	as *Bismarck*

Cost: RM196,800,000 (*Bismarck*), RM181,600,000 (*Tirpitz*)

Conclusion

The *Bismarck's* problems fall under two heads—inherent design flaws and tactical errors—which made her loss inevitable. The Germans were always very secretive about armour dispositions, but after the ship had sunk, rescuers found a set of flip-cards in the overall pocket of a petty officer. From this the Director of Naval Construction's staff reconstructed the internal layout of the *Bismarck*. This resolved another question of fundamental importance: why did this modern capital ship stop firing after only 20 minutes?

The *Bismarck* class, like the *Scharnhorst* class and the *Admiral Hipper* class heavy cruisers, suffered from weak stern structures. In the case of the *Bismarck*, remotely operated cameras examining her wreck on the seabed show that her stern was actually separated from the hull. Thus it is clear that the aerial torpedo-hit did not merely wreck the rudder; even without it the collapse of the stern would have made her very difficult to manoeuvre.

The June 1941 reconstruction of her internal layout astonished the DNC and his staff when they saw that the armoured deck was low down in the ship, as might be expected in a First World War design. In comparison, contemporary Royal Navy and US Navy battleships had their armour decks considerably higher to increase the volume of the protected citadel. Another feature of the design was the provision of a tertiary battery of 10.5cm anti-aircraft guns in addition to a heavy secondary armament of 15cm low-elevating guns in twin turrets. This three-layer armament, particularly the 15cm turrets, imposed a severe weight penalty. In the Royal Navy and the US Navy the decision had been made earlier to provide dual-purpose guns, a considerable saving in topweight and deck space.

German technical historians have recently discovered that the *Bismarck*'s 10.5cm guns were controlled by two different fire-control systems, one forward and one aft. To make matters worse, gun crews were not familiar with either system. The failure to destroy the Albacore and Swordfish attacks is often attributed to the aircraft speed being too low for the fire-control predictor's lower setting; the new evidence suggests that the fire control was not good enough.

A favourite question of the '*Bismarck* was unsinkable' school of thought is: was she sunk by gunfire, torpedoes or scuttling? An American expert once replied 'yes'! The evidence for scuttling rests on very dubious claims in the German media by people claiming to have been on board to the effect that the enginerooms were 'ready for the Admiral's inspection', and an order to fire the scuttling charges followed shortly afterwards. Unfortunately all these rather dubious claims ignore the testimony of survivors to the effect that the ship was an inferno between decks, and nobody from below was sighted after the action began. All the communications, electrical power, telemotor leads and boiler uptakes were above that low-armoured deck, and enemy gunfire had shredded everything except the main machinery. That would explain why no guns were firing after 20 minutes.

The claim that 'not a single shell penetrated the armour' is refuted by the underwater photographs, which showed some 400 holes in the hull. Eyewitnesses say that the *Bismarck* was a waterlogged hulk, and it is almost certain that in the final stages some shells went straight through the unarmoured portions of the hull, but something made those 400 holes, many of them heavy-calibre hits which penetrated the armour. The salvo of four 21in torpedoes fired by HMS *Dorsetshire* undoubtedly hastened her end, but she had already been flooded by thousands of tons of seawater.

Admiral Lütjens made three major tactical errors. The first was to leave the Baltic in daylight, making it very easy for the Royal Air Force to spot her. By far the biggest was to sail after a fuelling hose broke while she was taking on oil in Bergenfjord; the unfilled capacity accounted for nearly a third of her bunkerage. This error was exacerbated by the underwater hits from HMS *Prince of Wales*, which allowed seawater to contaminate a significant amount of the reminder. US Navy analysis corroborated the Royal Navy's opinion that Lütjens's biggest mistake was to start the Atlantic sortie after losing so much fuel. Apparently he believed the propaganda about her being unsinkable.

There is no denying that the *Bismarck* was a powerful battleship, but the *Kriegsmarine* does not appear to have spent the money to best advantage. Valid comparisons can be made with the US Navy's South Dakota class and the Royal Navy's suspended Lion class, displacing 40,000 tons, armed with nine 16in guns and protected by 15in belt armour and a thick armoured deck to match. The real problem is that the *Bismarck*'s qualities have been greatly inflated, largely for the reasons already mentioned. I myself have heard claims that Wotan Hard steel was used to make the side armour impenetrable. It was in fact a splinterproof structural steel that was used, with Wotan Soft being deployed where it was not necessary to stop shell-splinters.

The *Bismarck* met her end in a hail of gunfire, but she fought as long as she could and her officers and ratings died bravely. It is a pity that such bravery was not matched by good tactics and flaws in their ship's design.

Implacable class fleet aircraft carriers

Royal Navy 1939–1956

The story of the Royal Navy's armoured deck aircraft carriers goes back to 1936, when the Admiralty lobbied for an acceleration of the naval building programme. By now the threat from Nazi Germany and Fascist Italy could not be ignored by Parliament, and in June that year the Committee on Imperial Defence's sub-committee on Defence Policy & Requirements (DPR) asked for proposals. The First Lord of the Admiralty, the Navy's political head, suggested that the 1936 carrier programme should be expanded from one ship to two, and a month later a programme of two carriers in the 1936 programme, two in the 1937 programme and two in the 1938 programme.

The Controller, Rear Admiral Reginald Henderson, was the Navy's most senior aviator, and he exercised a very personal influence over the design of the new carriers. Fleet exercises showed that the threat from air attack was growing, and it was correctly assumed that carriers would suffer some damage. There were two ways of reducing the risk. One was to increase the complement of defending fighters, but to accommodate both strike aircraft (torpedo bombers or dive bombers) and defence aircraft in useful numbers made a large carrier mandatory. The alternative was to increase the carrier's resistance to damage, primarily by providing a powerful defensive armament.

The Royal Navy's carriers had a number of demanding tasks, and they would be operating within reach of shore-based bombers, so the powerful defensive armament was very sensible. But the threat from the gunfire of enemy cruisers was taken seriously by the three major aviation-minded navies. Reflecting this, both the US Navy and the Imperial Japanese Navy built carriers armed with 8in (20.3cm) guns. The British had provided their first-generation carriers with medium-calibre guns, but quickly lost interest in the idea, reasoning that the carrier's escorting cruisers were the right ships to neutralise enemy cruisers. There was, however, a perceived risk that enemy gunfire might wreck the hangar and flight deck, rendering the carrier unable to operate its aircraft, even if it was not sunk.

The problem was the reaction-time. Until radar came to the rescue after the outbreak of the Second World War, the first sighting of incoming bombers would be visual contact from one of the outer screen of destroyers. Aircraft were getting faster, cutting the reaction time even further. Henderson's answer was to provide the new carriers with armoured hangars, or more correctly, protected hangars, with flight decks armoured only over the length of the hangar, and lightly protected hangar sides. Such a scheme of protection carried significant weight-penalties. The

The launch of HMS Indefatigable *at the John Brown shipyard on the Clyde.*

weight of armour was high up in the ship's hull, where it had an inevitable effect on stability. The Royal Navy's insistence on a 'closed' hangar as part of a complex series of precautions against fire caused further problems by limiting aircraft stowage. Ton-for-ton Royal Navy carriers embarked fewer aircraft than their American and Japanese contemporaries with their 'open' hangars. The innovative carrier HMS *Ark Royal*, authorised in 1934, had been given a double-storey hangar, giving her a theoretical complement of 60 aircraft.

Henderson insisted that the 1936 Programme carriers should have protected hangars, but the Royal Navy was still shackled by international disarmament treaty limits on its total tonnage of carriers. The tonnage limit permitted the construction of five 27,000-ton carriers, but the Admiralty wanted six, so the standard displacement of the new carriers had to be held down to 23,000 tons. On this relatively modest figure it would only be possible to have a single hangar, nearly halving the complement of aircraft to a theoretical 36. This was the *Illustrious* class, the Royal Navy's latest design when the Second World War broke out in September 1939. It was originally planned to build six to the same design, but as things turned out the class was divided into three sub-groups. The *Illustrious*, *Victorious* and *Formidable* formed the first of these, while *Indomitable* was modified by the addition of a short lower hangar in an attempt to improve her capabilities. Basically, the improvements were those approved for the two ships of the third sub-group, *Implacable* and *Indefatigable*. The height of the hangar was increased by 6ft, and to compensate for the extra weight the side armour of the hangar was reduced from 4.5in to 1.5in. These modifications increased the aircraft complement to 45 aircraft.

At the end of the First World War the Royal Navy had the most advanced naval air force in the world, but already the rot had set in. On 1 April 1918, appropriately April Fool's Day, the

Royal Naval Air Service and the Royal Flying Corps were merged to form the world's first 'Independent Air Force', the Royal Air Force. The long-term effects were dire; air-minded naval officers transferred to the new force, and aircraft procurement remained firmly in Royal Air Force hands. As a result, the Royal Navy approached the international crisis in the early 1930s with markedly inferior aircraft. Even the belated decision to hand back responsibility for naval aviation in the mid-1930s had no immediate effect—the Royal Air Force took back all its maintainers and continued to have the lion's share of aircraft production facilities, including engines. Given the poor performance of the re-formed Fleet Air Arm's aircraft, the decision to emphasize their passive protection from bomber attack made sense, but the numbers and quality available at the outbreak of war were pitiful.

The war experience of the first four armoured carriers was impressive. In August 1940 HMS *Illustrious* was sent to join the Mediterranean Fleet as flagship of Rear Admiral Aircraft Carriers (Mediterranean). The Commander-in-Chief, Admiral Cunningham, needed no convincing of her value, and quickly put her to work to force the powerful Italian Navy on the defensive. On 11 November her Swordfish torpedo-bombers launched Operation 'Judgement', an attack on the main base at Taranto. In the first attack of its kind, the slow Swordfish sank the Italian battleship *Conte di Cavour* and seriously damaged the new battleship *Littorio* and the older *Caio Duilio*.

In addition *Illustrious*, with the old carrier *Eagle*, carried out a series of hit-and-run raids. She became such a threat to Axis naval operations in the Central Mediterranean that the *Luftwaffe* was ordered to sink her, using the specially-trained dive-bomber squadron *X.Fliegerkorps*. On 10 January 1941 the Ju 88 Stukas hit the carrier 75 miles east of Malta with eight 500kg and 1000kg bombs. Although seriously damaged, her combination of armour and fire precautions saved her and she crawled into Malta for emergency repairs before she heading across the Atlantic for total rebuilding at Norfolk Navy Yard.

Later, HMS *Formidable* and HMS *Indomitable* were hit by heavy bombs but survived. Their sister HMS *Victorious* played a major role in the pursuit of the battleship *Bismarck* in May 1941. All four were sent to the Far East in 1944–45 to take part in the final offensive against Japan.

The two ships of the third sub-group were suspended almost immediately to free industrial capacity (armour and shipyard capacity), but the delay permitted a major redesign incorporating war experience. This involved a return to a double-storey hangar and adoption of a more powerful four-shaft installation in place of the three-shaft arrangement in the four earlier carriers (boosting power by a third). The larger dimensions also permitted a more comprehensive air defence system, with more radar-controlled directors and close-range anti-aircraft guns. By the time the ships were well advanced, the Fleet Air Arm had adopted the US Navy's ideas of a deck park and outriggers as a means of expanding the aircraft capacity.

Up to this point all the changes to the design were beneficial, but now a series of problems arose. As planned, the lower hangar was to have a height of 16ft, while the upper hangar was be only 14ft, but it was decided to make both hangar heights 14ft. This decision took no account of the growing size of carrier aircraft, and the Royal Navy's biggest carriers built to date could not accommodate the Chance Vought F-4U Corsair, the aircraft which was to inflict unbearable attrition on the Japanese land-based and carrier air forces. The new ships could only accommodate the greatly inferior Seafire, a less than satisfactory adaptation of the land-based Spitfire, instead of the formidable 'bent-winged bastard from Connecticut'. HMS *Indefatigable* reached the Indian Ocean at the end of 1944 with an air group comprising 32 Seafire Mk IIIs for fleet defence, and 8 photo-reconnaissance Hellcats, 12 Fireflies and 21 Avenger torpedo-bombers, 73 aircraft in all. Her sister *Implacable* did not arrive until May 1945, with 48 Seafires, 12 Fireflies and 21 Avengers, 81 aircraft in all.

THE WORLD'S WORST WARSHIPS

Although the hull was slightly longer than the earlier sub-groups, it proved impossible to accommodate the additional personnel and storerooms required, and the forward part of the lower hangar had to be converted to mess decks. They were as a result no better off than the older *Indomitable*, with a short lower hangar served by the after lift. With hindsight one can only ask, was all the effort worthwhile?

After the war was over plans were drawn up to replace the two hangars with one 20ft high, but funds were not available. HMS *Implacable* was used for trials in 1946–48, and spent two years as Home Fleet Flagship, operating Sea Hornet twin-engined fighters and Firebrand strike aircraft. She was paid off for a cheap conversion to training duties, with accommodation and classrooms in her hangars, and was recommissioned in January 1952 in the Home Fleet Training Squadron. After less than two-and-a-half years in that role she was paid off for disposal in September 1954, and sold for scrapping the following year. Her sister HMS *Indefatigable* went straight into the training role and served in the Home Fleet Training Squadron from 1950 to 1954 before being sold for scrapping in 1956. The two of them had logged an average of less than seven years of active service each, including operations in Northern European waters, the East Indies and the Pacific.

The Implacable class were a victim of a mysterious decision to reduce the height of the larger hanger at a time when aircraft were getting larger. (Line drawing 1/2400 scale.)

IMPLACABLE CLASS

Implacable laid down 21 February 1939, launched 10 December 1942, completed 28 August 1944 by Fairfield Shipbuilding

Indefatigable laid down 3 November 1939, launched 8 December 1942, completed 3 May 1944 by John Brown

Displacement:	23,450 tons (standard), 32,110 tons (deep load)
Dimensions:	690ft (pp) 766ft 2in (oa) x 102 ft (max.) x 26ft 8in (mean)
Machinery:	4-shaft Parsons geared steam turbines, 148,000 shp; 8 Admiralty 3-drum boilers
Speed:	32kts
Armament:	16-4.5in Mk III AA guns (8 x 2), 52-2pdrs (6 x 8, 1 x 4), 16-20mm (7 x 2, 2 x 1)
Air Group:	72 (as designed)
Fuel/range:	4800 tons oil, 94,650 gallons avgas
Complement:	970 officers and ratings + 587 air group + 15 flag staff

CONCLUSION

The story of these two expensive but relatively ineffective aircraft carriers raises some interesting points. The first is their demonstration of the final outcome of the sad story of the decline of the Royal Navy's air arm from its high point in 1918. The 'dual control', whereby the Royal Air Force controlled the procurement and manning of aircraft but the Navy operated the carriers from which they flew, had been intended as a stopgap solution, but post-war the vociferous air power lobby preached a flawed doctrine of the 'obsolescence' of sea power as compared with

HMS Indefatigable *during the war. The Implacables lasted a very short time by comparison with other RAF carriers. both were scrapped by 1956, a mere twelve years after their completion.*

the 'cheapness' and radical alternative of air power. Out of this emerged the quasi-doctrine of the 'indivisibility of air power', which claimed that land-based bombers would replace both armies and navies and win strategic victories over cowed enemy governments. Great Britain's rulers and civilians were easily seduced by an argument which promised to avoid the mass slaughter of the Great War and to save money.

As already pointed out, the decision to hand control of naval aviation back to the Royal Navy was nearly too late to repair the damage done, and it is no exaggeration to say that the Fleet Air Arm went to war with woefully inferior aircraft. Miracles were achieved with the biplane Swordfish and Albacore torpedo-bombers, but even the newer aircraft types were unimpressive. Not until the US Navy came to the rescue in 1942, providing superb aircrew-training as well as high-performance fleet fighters and torpedo-bombers, did the Fleet Arm become an effective force.

Given the serious shortcomings of pre-war naval aircraft and the inability of the British aircraft industry to meet both Air Force and Navy demands on production of airframes and engines, Henderson and the First Sea Lord, Chatfield were right to emphasize passive protection for the new carriers. The heavy damage sustained by HMS Illustrious has already been mentioned, but later HMS Formidable was hit by two 1000kg bombs and HMS Indomitable was hit by two 500kg bombs; due to their protection, both were back in service within six months. In the intense operations around the Japanese Home Islands five out of the six were hit by kamikaze suicide strikes, but none was sunk or seriously damaged:

1 April 1945	*Indefatigable* hit at the base of the island
6 April 1945	*Illustrious* hit a glancing blow on the island
4 May 1945	*Indomitable* hit on after flight deck
	Formidable hit abreast of the island
9 May 1945	*Formidable* hit on flight deck (18 aircraft destroyed)
	Victorious hit twice

The late David Brown, former Head of the Naval Historical Branch and a specialist in carrier aviation, always insisted that however effective the armoured flight decks were, what really saved the British Pacific Fleet carriers from loss or serious damage was their excellent protection against petrol fires. In fact no British-designed carrier was lost in the Second World War from a fuel explosion. The closed hangar had its own ventilation and drainage systems; the first prevented avgas fumes from permeating the ship, and the second prevented water from fire hoses and sprinklers from causing the ship to capsize.

With the benefit of hindsight, it might have been better to build an improved (larger) Ark Royal design, but that would not have dealt with the problem of suitable aircraft. The long delays in revising the design of the *Implacable* and *Indefatigable* should have avoided the serious error of reducing the hangar height; aircraft were getting larger, not smaller. At this distance in time it is not clear why the decision was made, but it seriously reduced the value of the increase in dimensions and power. One of the less obvious lessons from the experience with the *Implacable* and *Indefatigable* is that an increase in size does not automatically result in an improvement of fighting power. Expanding the design of a complex warship like a carrier only works if all the interactions of sub-systems are carefully thought out. On the credit side, however, it must be noted that all six carriers could handle aircraft weighing 20,000lbs (9 tons), far in excess of any carrier aircraft weights at the time.

The protected hangar concept was right for the Royal Navy at the time, but it is interesting to note that the final wartime design, the 50,000-ton Malta class, would have had open hangars and a very large air group, like US Navy carriers. The air group of an aircraft carrier is its only offensive weapon, and a big air group is much better than a small one, for both fleet defence and strikes against the enemy. Logically, therefore, a carrier must always be designed around her air group; failure to remember that principle produces at best only

Hydrogen Peroxide Submarines

German Navy 1940–45
US Navy 1946–48
Royal Navy 1946–58
Soviet Navy 1947-59

Ever since the submarine became a practicable weapon of war designers and operators have battled with the problem of air-supply. The need to surface at frequent intervals to recharge batteries was a major limitation on the submarine's capability.

The earliest known example of an attempt to make a submarine independent of atmospheric oxygen was an experimental closed-cycle diesel engine tested in the Imperial Russian Navy's dockyard at St Petersburg in 1913. The Russian designers were also taking a step towards the snorkel: an air-mast which changed stale air in the boat and also reduced the risk of taking water down a hatch in rough weather. But it was not a snorkel as it was not used when running submerged. The air-mast was used in several submarines pre-war, but there is no record of the closed-cycle diesel being installed.

After Hitler repudiated the Versailles Treaty in 1935, the German Navy (initially the Weimar Republic's *Reichsmarine*, and later Hitler's *Kriegsmarine*) started to create a U-boat arm. The former U-boat officers advising on choice of design for the new generation were very aware of the risk to a submarine surfacing to charge her batteries, and looked for ways of extending underwater endurance. At the time the only option was the so-called Kreislauf closed-cycle diesel, and trials began; However, it did not meet the Kriegsmarine's requirements and it was never installed in an operational U-boat.

A more promising development was the Walter system, a turbine using an oxidant mixed with diesel fuel, producing enough heat to create a gas mixture to drive the turbine. In January 1940 the 85-ton experimental boat *V-80* was launched at Kiel, and started trials four months later. She was never anything but a test-platform, and was taken of service at the end of 1942. Three more experimental boats, *V-300*, *Wa-201*, *Wa-202* and *Wk-202*, were ordered in 1942, but only *Wa-201* and *Wk-202* were completed, and renumbered *U-792*, *U-793*, *U-794* and *U-795*. Commissioned in 1943-44, all four were scuttled in May 1945, but *U-792* and *U-793* were raised by a Royal Navy salvage team.

U-1407 *being examined by British officers in a German shipyard. The submarine was one of the Type XVIIB U-boats and after the war served in the Royal Navy as HMS Meteorite.*

The first production design was the Type XVIIB, of which twelve were ordered in 1942, followed by another twelve Types XVIIG of slightly improved design. Of these, only three were commissioned, the rest being scuttled before completion or cancelled. Although the fuel, enriched high-test peroxide (HTP, containing two atoms each of hydrogen and peroxide) or Perhydrol, was extremely unstable, and submerged endurance at speed was very low (123 nautical miles at 25kts), the Type XVIIB boats were fast, and might conceivably have helped Admiral Dönitz to overturn the Allies' 1943 victory in the Battle of the Atlantic if development had started two or more years earlier. As things turned out, they were far too expensive in scarce raw materials and too manpower-intensive, and by 1944 the Third Reich no longer had the resources to develop them. The bigger Type XVIII was suspended in March 1944, and the Type XXIV and the small Type XXVI were cancelled.

TYPE XVIIB U-BOATS

Displacement:	312 tons (surfaced)
Dimensions:	41.45m (oa) x 4.50m x 4.30m
Machinery:	Single-shaft Walter HTP turbine, 5000shp, 8-cylinder supercharged diesel, 210 bhp
Speed:	25kts on Walter turbine/7.5kts on electric motor
Armament:	2-533mm bow torpedo tubes (+ 2 reloads)
Endurance:	123nm @ 25kts; 3000nm @ 8kts on diesel
Electronics:	sonar?
Complement:	19

This was not the end of the ideas promoted by Dr Hellmuth Walter. *U-1406* was raised and given to the US Navy for trials, running until 1948. U-1407 ran as HMS *Meteorite* from 1946 to 1949, providing data for the Royal Navy's own Walter programme. She was described officially as '75 per cent safe', but she provided useful experience for two prototypes intended to be followed by an operational class of twelve boats. HMS *Explorer* and HMS *Excalibur* (completed in 1956 and 1958) were for a while the fastest submarines in the world, exceeding 25kts, but one officer described them as 'not for the fainthearted' and they were known as 'Exploder' and 'Excruciator' on account of the frequent problems with the fuel. HTP burns through cloth, flesh and bone, and something as small as a speck of dust or a flake of rust can cause it to overheat and ultimately explode. As the fire was feeding on its own oxygen, a foam blanket did not work; the only way to extinguish a fire was to drench the open fuel tank with seawater to bring the temperature back to a safe level.

Explorer class

Displacement:	980 tons/1076 tons (surfaced/submerged)
Dimensions:	225ft (oa) x 15ft 8in x 18ft 2in
Machinery:	2-shaft Vickers HTP turbine, 15,000 shp. Electric motor, 400hp
Speed:	27kts on turbine; 18kts on electric motor
Armament:	none
Endurance:	not available
Electronics:	187 sonar
Complement:	41–49

HMS Excalibur *in April 1962. The ship was known more familiarly in the Royal Navy as the* Excruciator.

In parallel, the Soviet Union started experiments before the Second World War, theoretically putting the Red Navy well ahead of the Germans, developing the Kreislauf system, in which liquid oxygen (LOX) was used as an oxidant. Information on these developments has many gaps, but the first trials were probably run by *M-92*. One unit, *M-401*, was powered by an improved *Regenerativny Yedinyi Dvigatiel Osobovo Naznacheniya* (REDO) plant. The *Kriegsmarine* submariners had been right: the Kreislauf system was inherently dangerous. Several met with serious accidents during the Great Patriotic War of 1941–45 and post-war.

The advancing Red Army captured large quantities of submarine material as it advanced along the southern coast of the Baltic, ranging from whole sections of boats down to Walter powerplants, snorkels and advanced torpedoes. This outflanked any attempt by the Western Allies to keep this technology out of Soviet hands.

The first visible result was the appearance of the Project 615 coastal design, known to NATO as the 'Quebec' class; it was chosen instead of the conventionally-powered Project 612. The known programmes suggest that 33 entered service between 1950 and 1962; they were 460-ton boats armed with 4–21in torpedo-tubes forward. Their powerplant was unusual; a triple-shaft arrangement with a Kreislauf system driving all three shafts for submerged running and a back-up 900 bhp diesel engine on the centre shaft and 700 bhp diesels coupled to the outer shafts.

The Kreislauf plant ran on LOX, added after the exhaust had been passed through a lime-based chemical absorbent, known in the Red Navy as ED KhPI.

Because LOX can be stored only for short periods, the 'Quebec' class never operated far from their bases. To remedy this, a Project 637 design was drawn up, using a plant designated B-2. In this system solid granules held bonded oxygen and a carbon dioxide absorbent; exhaust gas was passed through this section to 'scrub' the carbon dioxide and it was recharged with fresh oxygen. While under construction at the Sudomekh yard M-361 was converted in 1958-59 to a B-2 test-bed, but the whole programme was terminated when the design bureaux succeeded in producing a workable pressurised water reactor plant.

Despite attempts to improve the resistance to fire and explosion in the later boats the Kreislauf system remained dangerous; known accidents occurred to: *M-401*, *M-255*, *M-256* (sunk after an explosion in 1957), *M-257*, *M-259*, *M-351* and *M-352*.

When the loot from Germany was looked at after the surrender, the Soviet designers came up with Project 616, a Type XXVI Walter-engined U-boat completed with equipment seized at the same time. It was soon obvious that such a 'lash-up' was not going to work, and 616 gave way to Project 617, which would use the same type of 7500 shp Walter plant in a hull built to Soviet shipyard practice. The preliminary design was completed by the end of 1947, and German Walter engine specialists helped to build and test the new powerplant at a land test-site until 1951, when construction started on the prototype *S-99* at Sudomekh. She was launched in 1952 and commissioned in June the same year.

German practice was followed closely, with HTP stored in plastic bags in tanks slung below the main pressure hull. Starting the Walter plant took only two minutes, and from a cold start to full speed (20 knots) took about nine-and-a-half minutes. The boat lost power as it dived deeper, because of the need to dump exhaust gas overboard. On trials the engine was run at a maximum depth of 260ft; at a depth of 95–130ft the powerplant produced 6050 shp. Soviet records claim that *S-99* went to sea 98 times from 1956 to 1959, a total of 6000 nautical miles on the surface and about 800 nautical miles submerged. The four-year gap between commissioning and that series of sorties is puzzling; perhaps the time was taken up with harbour trials and training.

On 19 May 1959 *S-99* suffered an explosion while trying to start the Walter plant at a depth of 260ft. Mud had apparently clogged the valve of the HTP supply-pipe; the HTP decomposed and ultimately exploded, blowing an 3in hole in the pressure hull. Although *S-99* was raised she was never repaired. As with the Project 615 'Quebecs', nuclear power was a much more attractive option, and research into Walter technology and other ideas stopped.

Particulars of Project 617

Displacement:	950 tons (surfaced)
Dimensions:	204ft x 19ft 11in x 16ft 7in the Type XX
Machinery:	Single-shaft Walter turbine, 7250 shp 26-cell battery, electric motor, 540 hp 'creep motor' 140 hp 8-cylinder 4-stroke auxiliary diesel, 450 bhp
Armament:	6-533m bow torpedo tubes (+ 6 reloads)
Speed:	20kts (submerged) on Walter plant, 9.3kts on electric motor
Range:	120nm @ 20kts on Walter plant; 134nm @ 9.3kts on electric motor
Electronics:	Nakat radar; Tamir-5 and Mars-24 sonars
Complement:	51

Conclusion

As we know, the Walter turbine powerplant was to prove a blind alley, like the Kreislauf engine before it. For the Kriegsmarine, the pursuit of a solution became more urgent as the Third Reich crumbled under the Allies' attacks. Air-independent propulsion changed from giving an edge in attacks on shipping into a vital means of surviving enemy anti-submarine measures. In that sense, the Walter system was complementary to the snorkel air-mast.

The Third Reich's choice of priorities was always guided by prejudice rather than logic, and under the pressures of total war, the allocation of resources depended on who had the ear of the Führer. The Walter programme consumed scarce commodities at an alarming pace, notably skilled labour. It is typical of the way the war economy was run that the radically more advanced Type XXI 'electro U-boat' suffered delays because the most skilled workers were working on the outdated Type VII design and the various Walter designs.

Post-war, both the US Navy and the Royal Navy found the cost of the fuel a major drawback. A US Navy engineer calculated that Perhydrol was 80 times as expensive as diesel fuel. The nature of HTP also dictated elaborate support measures. For example, the crews of HMS *Explorer* and HMS *Excalibur* never 'owned a plank', and were accommodated in a tender.

The teams setting *U-1406* to work in the US and their opposite numbers in the UK doing the same for *U-1407* (HMS *Meteorite*) soon found that the claims for the Walter system, and for other recent designs like the Type XXI, were exaggerated. In December 1946 Captain Logan McKee wrote that the design of the turbine was needlessly complex, and suggested that German designers preferred to play safe rather than risk being sent to the Eastern Front, or worse. In 1947 the Admiralty published a secret report claiming that the designs of both the Walter boats and the Type XXI had inherent flaws. It was concluded that the increase in diving-depth was achieved at the cost of major reductions in safety-margins. We now know that the streamlined hulls of the Type XXI and Type XVII were not tank-tested to verify the hull-forms, and as a result may have offered more hull drag than necessary. Another weakness was their relatively shallow diving-depth. HMS *Meteorite* was limited to a maximum depth of 100ft.

The US Navy lost interest in closed-cycle machinery because it knew that nuclear propulsion was not only feasible but also developing fast. The Royal Navy was feeling the pinch because of a national financial crisis, and also saw its nuclear propulsion programme slip rapidly down the priority list.

What, then were the benefits of HTP-fuelled submarines? HMS *Explorer* and HMS *Excalibur* certainly contributed to the search for high underwater speed and played a useful role in the development of new anti-submarine tactics. The technology may have been ahead of its time, but the submarine world is still searching for the perfect air-independent propulsion (AIP) system.

Today the danger to 'conventional' (non-nuclear) submarines from hostile aircraft—helicopters, surface ships and long-range passive detection—is so great that AIP is seen as essential for the survival of shallow-running submarines. The Kockums division of the Swedish Karlskronavarvet shipbuilding combine has pioneered the Stirling engine, using LOX and diesel fuel. The Royal Swedish Navy has seven Stirling-engined boats in commission and Japan's Maritime Self Defence Force has just converted the *Asashio* by inserting an additional hull-section for a Stirling Mk 2 engine. The German Navy prefers the fuel cell, which uses a chemical reaction to convert heat into electrical power. It requires very expensive components, but has the important advantage of not requiring LOX to be carried on board; In addition to the *Deutschemarine*, the system has been sold to Italy and Greece. Russia's Rubin Bureau offers its own fuel cell-based AIP, but there is no evidence suggesting that it has been installed in any submarine. The

The destruction of HMS Sidon illustrated the dangers inherent in HTP.

French Navy never bought the MESMA plant, which uses a turbine driven by LOX and methanol, but has sold it to Pakistan.

The renewed interest in LOX is nurtured by advances in its handling pioneered by the offshore oil industry. In the past, navies have dismissed civil-sector technology because it is insufficiently 'rugged', but nothing is more rugged than the environment of a drilling platform!

HTP never really went away. The US Navy, the Royal Navy and the Red Navy were all impressed by the advanced torpedoes found in Germany in May 1945. The Royal Navy's 'Fancy' Mk 12 had its Russian and US equivalents but its operational life came to an abrupt end in 1955, when a Mk 12 being loaded aboard the submarine HMS *Sidon* exploded and sank her with heavy loss of life. HTP was eventually identified as having caused the disaster, and the Mk 12 was soon withdrawn from service.

When the Royal Swedish Navy was looking for technical assistance in developing high-speed torpedoes it asked the British, who sold them not only the design of the Mk 12 and all the research that had gone into it, but also the detailed enquiry results, all for a nominal sum. The Swedish torpedo-manufacturer FFV was given the task of sorting out the problems. Their findings were that HTP is so sensitive that it needs a specially designed fuel system, something that the British had not done. The result was the Tp 61 series of weapons, capable of very high speed (60kts+), but distinguished by high reliability and safety. The Soviet Navy managed to get away with no (publicly admitted) accidents until August 2000, when a torpedo exploded in one of the tubes of the big submarine Kursk, triggering off a massive explosion as the reload torpedoes caught fire and blew up only seconds later, killing all 118 on board.

'ALPHA' CLASS
NUCLEAR ATTACK SUBMARINES

SOVIET NAVY 1961–95

The Soviet Navy's interest in nuclear-powered submarines was spurred on by the commissioning of the USS *Nautilus* (SSN-571) in 1955, although the concept was in embryo before that. In fact recent Soviet engineers' memoirs claim that Stalin's henchman Lavrenti Beria deliberately delayed the programme because he did not understand the technology! Under a September 1952 directive separate design groups were established to develop a strategic nuclear-powered submarine and its powerplant. Formal design was begun in March 1953 by the SKB-143 bureau (later named the Malakhit Bureau). After much chopping and changing the design emerged as a nuclear attack submarine (SSN), armed with eight 533mm bow torpedo-tubes (20 torpedoes carried). Displacement was 3050 tons surfaced, 4750 tons submerged, and the powerplant was a pair of VM-A reactors driving twin shafts and diesel generators developing 33,000 hp. In addition two 450 hp 'creep' motors were provided for slow, silent running. Speed was 24kts, diving-depth 300m, and the boats were stored for 50 days.

The first Project 627 SSN was named *Leninskiy Komsomol* (K-3), and the class, named 'November' ran to 12 units. K-3 started trials in July 1958, only three years after the *Nautilus*, but the class had a chequered history. K-3 had steam-generator trouble on her way to the North Pole in 1962, suffered a fire in 1967 and was taken out of service. K-5 had a serious reactor accident, as did K-11, and K-8 sank off Cape Finisterre after a major fire in 1970. Although very noisy, their performance surprised the US Navy, and led to pressure for building the Los Angeles (SSN-688) class. The VM-A plant was used for the Project 658 'Hotel' and Project 659 'Echo' classes, hence the designation HEN by Western intelligence ('Hotel', 'Echo', 'November').

The Soviet military saw the need for submarines capable of launching strategic (i.e. land attack) missiles against American cities, as a straightforward means of offsetting the United States's military superiority. After some particularly dangerous experiments with liquid-fuelled missiles fired from diesel-electric boats (SSGs), attention turned to a ballistic missile-armed derivative of the 'November' design, the Project 658 'Hotel' SSBN. This was followed by the Project 659 'Echo' SSBN, the second group of which were armed with long-range tactical cruise missiles, turning them into SSGNs. Their targets were US Navy aircraft carriers, which emerged as a new threat because they could launch nuclear strikes against the flanks of the Russian landmass.

The Project 667 'Yankee' design was the first Soviet SSBN to resemble the US Navy's George Washington design, leading to speculation that the KGB or the GRU military intelli-

gence organization had stolen the plans. But the resemblance was only superficial, and the two designs were very different indeed. These SSBNs were followed by the Project 667B Murena class ('Delta I') design, which went through another three step-improvements in armament. The development of SSGNs continued apace, with the Project 670 Skat class ('Charlie') series and the one-off Project 661 Anchar ('Papa'). Another series of SSNs, the Project 671 Ersh class ('Victor II') began production in 1965. The Soviet Navy published no audited accounts and there was not even a published defence budget (such are the joys of a command economy in a dictatorship), so we cannot put figures on the total cost of all this submarine-building. It must have been a colossal drain on the defence budget, however, even using a rough comparison with Western research and development, building and running costs.

The Project 705 *Lira* ('Alpha' class) nuclear attack submarine (SSN) was conceived in 1958 by A B Petrov, an engineer working for the Malakhit Bureau. According to one Soviet source, the original requirement was for a 1500-ton SSN capable of 40kts, but the design grew to 2300 tons. Petrov aimed at a hull displacing half that of the latest SSNs, to be achieved by having a single hull. Other weight-reductions were to be achieved by using titanium instead of high-tensile steel, adopting automation to cut the crew to 15–17 men, and adopting a small light gas- or liquid-metal cooled reactor.

The project had the backing of V N Peregodov, head of Malakhit, and he created a special group under Petrov within the Bureau to develop the design. The Soviet Navy's submariners did not like the idea of a single reactor and a single hull, but Admiral Gorshkov, Chief of the Soviet Navy, and B Y Butoma, Chairman of the State Committee for the Shipbuilding Industry, both supported the project. The design-study was completed in June 1960.

Butoma wanted to see the project completed, but he replaced Petrov with the chief designer M G Rusanov in May 1960. The project was finally authorised on 23 June that year and the design concept was completed on 31 December. Along the way Petrov's radical ideas had been diluted; the final design had a double hull, six compartments rather than three, double the crew and double the displacement. What survived, however, was the *Akkord* automated combat information system, which allowed all functions to be run from a single console. The *Lira* was thus the first Soviet submarine in which all electronics, including the navigation system, surface search radar, sonar system and fire control were controlled from a single central console. The torpedoes were ejected hydraulically, allowing them to be launched at any operational depth. Another innovation was the incorporation of an escape capsule into the fin. Very careful attention to hull-shape, enclosing all masts in the fin and a slotted rudder, conformed to Petrov's goal of very high speed and exceptional manoeuvrability.

To reduce hull-weight by a significant factor titanium was chosen for the double hull. This metal exists in large quantities in Russia, but it is notoriously difficult to work with. At vast expense the Soviet Navy had built a special plant to machine titanium sections for the Project 661 *Anchar* prototype. As this ultra-fast SSGN is seen by reputable Western analysts as a missile-firing equivalent to Project 705, it would make sense to use the expensive titanium techniques for the smaller SSNs. Just to confuse the picture, however, recent research in the United States suggests that some of the Project 705/705K boats had conventional high-tensile steel hulls.

Two liquid-metal (lead-bismuth) powerplants were already under development in parallel. The VM-40/A was a modular two-section system with two steam lines and circulating pumps. OK-550 was also modular, but had branched first-loop lines and three steam lines and circulating pumps. VM-40/A was selected for the Project 705 boats built in St Petersburg, but OK-550 went into the Project 705K boats built at Severodvinsk. To save weight the standard 500Hz electrical power system in earlier Soviet designs was replaced by 400Hz systems in both groups.

Opposite: A 'November' class submarine, the first nuclear attack submarine in the Red Navy, surfaces from being submerged. The class suffered from several accidents involving their reactors.

Known to the Soviet Navy as 'Golden Carps' on account of their colossal cost, the enthusiasm for these SSNs was on the wane by 1973. Central Committee Secretary for Defence Sectors Dimitri Ustinov denounced their cramped design and impossibility to repair. The first boat in service, *K-377*, had suffered a reactor accident, reputed to be the result of the liquid-metal coolant freezing, and was scrapped in 1974. In April 1982 *K-123* lost her primary loop through a meltdown in 1982 and took nine years to repair. Butoma, now Minister of the Shipbuilding Industry, called for cancellation, and Rusanov was sacked. However, the six surviving units were modernised by 1982. A major problem which emerged was the lack of suitable bases. The primary reactor loop, the lead-bismuth section, had to be kept hot to keep the alloy liquid when the reactor was switched off for maintenance. It was also necessary to prevent the liquid alloy from oxidizing by re-generating it periodically, and to monitor its state constantly. They proved unsuited to long patrols, on account of the cramped hull, and the machinery was never 100 per cent reliable. One of the class was reported to be scrapped in 1988, and another was reported to have been recommissioned in 1989 after a five-year refit.

After the loss of the experimental Project 685 *Plavnik* type SSN (better known as *K-278 Komsomolets* or 'Mike' to NATO) in 1989, the Naval Staff decided to withdraw the Project 705 boats. Their powerplants were regarded as potentially unsafe, despite the fact that none had seen much use. At least one, possibly *K-123*, had her liquid-metal cooled system replaced by a pressurized water reactor plant and redesignated Project 705ZhMT.

News of the new high-speed design reached the West around 1971, apparently when one was tracked by a US Navy SSN. Soon an 'Alpha' Scare became public knowledge, with wild speculation about a new generation of 50-knot Soviet SSNs. It was widely assumed that all future SSNs would be derived from the 'Alpha', with great diving depth and high speed. Very expensive weapons were developed to match its performance, notably the US Navy's Mk 48 Mod 5 ADCAP (ADvanced CAPability) and the Royal Navy's Spearfish submarine-launched heavyweight torpedoes. Both were designed to run deep at 55 knots, a considerable technical achievement. In addition new lightweight torpedoes were developed for launching by surface warships, helicopters and maritime patrol aircraft, notably the US Navy's Mk 50 Barracuda and the Royal Navy's Stingray. These 324mm (12.75in) lightweights used shaped-charge warheads to attack the most vulnerable point in a Soviet submarine's double hull (the control room). In practice this meant designing the logic of the torpedo's seeker to make a terminal manoeuvre which brought it to a 90 degree angle to the hull at a pre-selected point in its length.

'ALPHA' CLASS SUBMARINES

K-377 launched 1967★ and delivered in 1972★ for trials before being dismantled, built by Admiralty Yard, Leningrad

K-316 launched 1974★, completed 1979★, built by Admiralty Yard, Leningrad

K-373 launched 1976★, built by Admiralty Yard, Leningrad

K-123 completed 26 December 1977, built by Severodvinsk Shipyard

K-432 launched 1978★, completed 1983★, built by Severodvinsk Shipyard

K-463 launched 1980★, built by Admiralty Yard, Leningrad

K-493 launched 1981★, completed 1983★, built by Severodvinsk Shipyard

★ speculative dates

Displacement:	2310 tons (surfaced), 4320 tons (submerged)
Dimensions:	267ft 1in x 44ft 3in (max), 31ft 2in (pressure hull) x 22ft 8in
Machinery:	1-shaft liquid-metal VM-40/A or OK-550 nuclear reactor; 38,000 shp

THE WORLD'S WORST WARSHIPS

The 'Alpha' class were an attempt to revolutionise the nuclear attack submarine. However, instead of a revolutionary step forward, the Red Navy created an evolutionary dead end.

	driving 1 OK-7 geared steam turbine; 2 136 hp 'creep' motors; 1 auxiliary diesel
Speed:	43kts (max); 14kts (quiet running)
Armament:	6-533mm torpedo tubes (bow); 18 torpedoes or mines
Diving depth:	350m (normal); 420m (max)
Patrol endurance:	50 days
Complement:	30 officers

CONCLUSION

Although the cost of the Lira class SSNs has never been revealed, circumstantial evidence suggests that they must be the most expensive attack submarines of their generation. The nickname 'Golden Carp' is known to be a reference to their cost of construction and through-life maintenance, even if the research and development costs are ignored. Their sudden fall from grace in the early 1970s suggests that even the prodigal Soviet Defence Ministry was uneasy about their cost-effectiveness.

The immense resources devoted to the Project 705 boats might have been worthwhile if they had given the Soviet submarine force a decisive edge in the Cold War. But they were clearly not worth the money spent on them, because they saw so little service. Their only significant achievement seems to have been to scare the living daylights out of NATO and Western navies in general. A cynical view might be that they were immensely valuable to the West because they contributed to the eventual bankruptcy of the Soviet Union and also sounded a wake-up call to the US Navy and the Royal Navy. An arms race based on technology was bound to favour the West, despite 'croakers' in the West who always credited the Soviets with being 25 light-years ahead, thanks to their ideological approach. A whole industry grew up in the West dedicated to exaggerating the Soviet threat, predicated on the perfection of technology not yet glimpsed in the West.

It has often been said by Western commentators that cost was never a factor in Soviet warship design, because the Navy's budget was never subjected to public scrutiny. But resources are finite, and we know from senior Soviet admirals' testimony that defence was consuming 15 per cent of gross domestic product (GDP), without taking account of infrastructure costs. By that criterion the Project 705 was only one of a number of straws which eventually broke the camel's back. It seems that the impetus for Project 705 arose from a desire to 'leapfrog' development by ten years, to guarantee technical superiority over the US Navy's SSN force. The tactical mission seems to have been almost an afterthought, and to this day nobody agrees on exactly how they were to be used. One plausible theory is that they were intended to act as 'interceptors' remaining in harbour or nearby, ready to respond to a call for assistance from surface ships or aircraft which had detected an enemy submarine. Another view was that their main targets were US Navy aircraft carriers, but the Russians have never made this clear. Perhaps there was no specific rationale for them; hard to believe, except in the Soviet context.

One of the many problems encountered in running the 'Alphas' was the relatively crude state of automation available to Soviet designers. As the automated control system was a basic feature of the design it guaranteed problems. Nor was Soviet industry good at designing integrated combat systems, because it had not kept pace with Western electronics developments. This lamentable state of affairs has been identified as the result of that self-same command economy which was supposed to ensure a massive advantage over the West. It was made worse by the corruption and nepotism endemic in the Soviet system, which often resulted in inferior equipment being supplied to the Armed Forces simply because the head of the factory supplying the equipment had sufficient political influence to deflect criticism.

Given the poor standards of safety designed into Soviet military equipment, the fact that the Project 705 and 705K boats were regarded as unsafe by the Soviet authorities themselves suggests that they were very dangerous. Soviet standards of health and safety are reckoned to have been about a century behind the West, largely because expectations were low, but partly because of patriotism. One of the 'Hotel' class SSNs was nicknamed *Hiroshima* by her crew!

The saga of Project 705 is an excellent example of a fundamental problem which preceded the Revolution, and seems to be still prevalent in Russia today. Nobody can deny that Russian scientists and engineers have produced brilliant innovations and solutions, but the execution of the ideas has all too frequently been so poor or haphazard that the nation has never reaped the full benefit. It also shows that ultra-rapid revolutionary development seldom results in success. Ships, especially nuclear submarines, are expensive capital items which cannot be discarded as soon as a design-fault reveals itself. Admiral Gorshkov, hailed by many idolaters as the father of modern Soviet seapower, fell into the same trap as Grossadmiral Tirpitz did in the years leading up to 1914. Given his head, he approved design after design, hoping to outbuild the Americans, with the result that the Americans took up the challenge. President Reagan said: 'They can start an arms race if they like, but we will finish it.'

Type 21 anti-submarine frigates

Royal Navy 1964–

In the mid-1960s the Royal Navy was looking for replacements for four diesel-engined frigates. This requirement coincided with vociferous but ill-informed criticism of the current *Leander* class frigates, some it from naval officers and some of it from pundits and the media. The agitation was seized on by the long-established frigate-builders Vosper Thornycroft as a business opportunity, and an intense lobbying offensive was launched. The theme of the campaign was the alleged conservatism of the Navy's designers in the Ship Department at Bath (successors to the DNC's department). The media was repeatedly briefed on the failings of the 'old women of Bath', the RCNC, who refused to countenance alloy superstructures.

There is nothing inherently dangerous about marinised aluminium alloy, and it was used at the time in the US Navy's large missile-armed destroyers, to offset the amount of topweight created by radars and trackers. The marinised alloy contains magnesium, however, making it inflammable at high temperatures, and it also likely to break up into fragments when exposed to blast. For that reason the Director-General Ships' (D-G Ships) department always insisted on steel superstructures. This imposed penalties of topweight, and some private shipbuilders were critical of the practice; the Royal Navy was at the time under pressure to keep displacement down, under the mistaken assumption that an increase in dimensions had a direct impact on cost. In its propaganda Vosper Thornycroft made the claim that a frigate designed to commercial standards could carry a more powerful armament than the officially designed *Leander* class. Assuming a baseline cost of £5 million for a *Leander*, it was claimed by the shipyard that it could build such a frigate for only £3.5 million, a difference of 40 per cent.

The argument was very persuasive to a Royal Navy short of funds for building new frigates at a time of inflation and spiralling costs. The government of the day announced that it would order four Type 21 frigates to a joint Vosper Thornycroft and Yarrow design to incorporate the best of commercial practice. In a moment of enthusiasm, motivated by the need to create work for the shipbuilding industry, the Cabinet instructed the Ministry of Defence (MoD) to double the order to eight ships.

So far so good, but the shipyards were given their head by the Controller of the Navy, who had little sympathy for D-G Ships and his team. They were told that they had complete freedom from interference from Bath, making nonsense of the claim that the Type 21 programme would use the best ideas of both teams. Years later the D-G Ships of the day admitted to the author that this decision resulted in a missed opportunity for both sides to benefit. He also claimed that the Type 21 design was deficient in stability, and insisted that the beam was increased. The two shipyards pooh-poohed his fears but D-G Ships threatened to go public unless they met his department's objections. There were other less crucial objections, but they were all brushed aside.

When the lead ship HMS *Amazon* was accepted by the Navy she attracted much interest, but the builders made the preposterous claim that she had exceeded 34kts on trials. As she was propelled by the new standard outfit of Olympus gas turbines for main drive and a pair of the smaller Tyne gas turbines for cruising, her maximum speed cannot have exceeded 30kts on full power and 18kts on the Tynes alone. Even more startling was the fact that her cost was £14.4 million, an increase of 400 per cent over the cost promised by the builders. Yet, such was the enthusiasm for the 'radical' new design that the builders were never called to account for the staggering rise in cost. The first three contracts were awarded to Vosper Thornycroft, including lead-yard services, and the remaining five contracts went to Yarrow Shipbuilders.

There were several good features in the new frigates, notably an improved internal layout and the substitution of the new 4.5in Mk 8 gun in place of the old 4.5in Mk 6 twin mounting. However, the new gun had already been funded by the Navy, so it hardly counted as an innovation by the designers. The officers' accommodation was better than in previous ships, but the ratings' accommodation was inferior, a point which was lost on the officers who sang the praises of the ships of the class. As so often happens, the officers who serve in warships are by no means objective in their judgements, and none of them questioned the use of alloy, let alone the cost. Nor did anyone in the fleet complain about the fact that they were designed for short hull-lives: 22 years rather than the 25–30 years expected of more robust ships like the *Leanders*. Far from stimulating exports, the Type 21 did not prove sufficiently attractive to Argentina and Australia, largely because they were expensive, an indictment of the whole policy of delegating design-responsibility to industry.

The Type 21 ships remained popular with the seagoing fleet, but at Bath there were misgivings about their light construction, and when the Falklands War broke out in April 1982 private fears were expressed that the severe weather conditions in the South Atlantic would inflict serious structural damage.

These fears had already been expressed during the so-called 'Cod War' off Iceland, but they were not made public. Another problem came to light when modernisation was considered in the early 1970s. There was a need for ships to operate against Soviet submarines transiting the Greenland-Iceland-UK (GIUK) Gap, using the new Type 2016 long-range sonar, and Seawolf missiles to meet the threat from anti-ship missiles fired from 'Charlie' type submarines. Although five *Leanders* were rebuilt to meet the new requirement, the Type 21s were found to lack stability for a similar conversion; they could have either the 2016 long-range sonar or Seawolf, but not both. As a result they never received any major upgrades while serving with the Royal Navy, and were earmarked for early disposal.

The outbreak of the Falklands War early in April 1982 resulted in seven Type 21s going south: *Active*, *Alacrity*, *Ambuscade*, *Antelope*, *Ardent*, *Arrow* and *Avenger*. They were about to face the ultimate test of a warship design, a real war in an extremely tough environment.

With no airborne early warning (AEW) to safeguard the Operation 'Corporate' task force in the confined waters off the landing beach, the Royal Navy had to do what it could to defend the beachhead from daily air raids by aircraft of the *Fuerza Aérea* and the *Armada Republica Argentina*. Some of the Type 21s, with their obsolescent Seacat short-range air defence missile systems, were stationed on the 'gun line' in an effort to break up the attacks and to provide gunfire support for the troops ashore.

When the destroyer HMS *Sheffield* was disabled by a hit from an air-launched Exocet missile on 4 May the *Arrow* went alongside to help with firefighting, directing hoses on the destroyer's side. The *Active*, *Arrow* and *Avenger* were with the second wave of reinforcements sent south in mid-May. The *Alacrity* made a daring night reconnaissance run through Falkland Sound, driving the enemy transport *Isla dos Estados* ashore and destroying her cargo of aviation fuel and army vehicles. The *Ardent* provided gunfire support during the first landings of British troops at San Carlos on East Falkland, the principal island of the group, but was hit by seven bombs dropped by Skyhawks on 21 May, and had to be abandoned; she sank the following day.

On 23 May the newly arrived *Antelope* was attacked by Skyhawks and was hit by a bomb which did not explode. It came to rest just over the bulkhead separating the two engine rooms.

The problems of the Amazon class were apparent from early in their careers. In 1977, a fire aboard HMS Amazon *called into question the use of aluminum in the superstructure. In 1993 all the surviving members of the class were sold to Pakistan. HMS* Alacrity *(below) was renamed* Badr.

THE WORLD'S WORST WARSHIPS

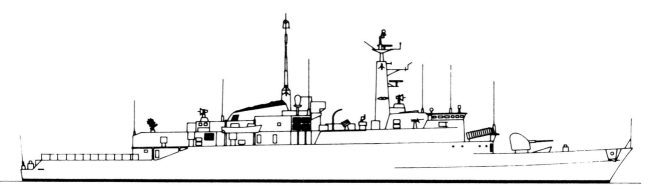

HMS Alacrity *in 1980, one of a Type 21 class that departed from many traditional Royal Navy building principles. (Line drawing 1/720 scale.)*

A second attack scored another hit, this time in the forward part of the ship. In the words of her commanding officer, 'it rattled around the front of the ship' before coming to rest in a cabin. After repelling another attack HMS *Antelope* was given permission to move to a more sheltered anchorage, where repairs could be started, and a bomb disposal team could be sent to deal with the two unexploded bombs. Unfortunately the attempt to defuse the first bomb was unsuccessful, and a huge explosion killed the Royal Engineer sergeant leading the disposal team, injured his assistant seriously and caused lesser injuries to two naval engineers helping. The ship was engulfed in flames, and firefighting was virtually useless because the blast had severed the fire main. The crew was taken off without further loss, leaving *Antelope* burning in the dark, until about midnight, when she was destroyed by a massive explosion in the Seacat magazine.

The *Ambuscade* and *Avenger* both survived Exocet attacks, proving that their passive countermeasures were effective when the correct procedures were followed. In fact the attack on *Avenger* was made by the last air-launched Exocet the Argentines possessed, but *Avenger* was later attacked unsuccessfully by a land-based version.

With the end of the war the surviving units returned to Britain. As predicted, they were found to have serious cracks in the upper deck and superstructure, and required stiffening. Some work was done to reduce vibration and to reduce machinery noise, but the days of the six survivors were numbered. In 1993–94 they were sold to Pakistan, and even that transaction proved controversial. The Defence Export Sales Organisation (DESO) was attacked for getting rid of the ships without inviting British companies to bid for the lucrative job of modernising and upgrading them. As a result of DESO's haste to get the ships off their books, they underwent a major modernisation with equipment and weaponry from a variety of European suppliers.

Type 21 Frigates

Amazon laid down 6 November 1969, launched 26 April 1971, completed 11 May 1974
Antelope laid down 23 March 1971, launched 16 March 1972, completed 19 July 1975
Active laid down 23 July 1971, launched 23 November 1972, completed 17 June 1977
Ambuscade laid down 1 September 1971, launched 18 January 1973, completed 5 September 1975
Arrow laid down 28 September 1972, launched 5 February 1974, completed 29 July 1976
Alacrity laid down 5 March 1973, launched 18 September 1974, completed 2 July 1977
Ardent laid down 26 February 1974, launched 9 May 1975, completed 13 October 1977
Avenger laid down 30 October 1974, launched 20 November 1975, completed 15 April 1978

Displacement: 3100 tons (normal); 3600 tons (deep load)
Dimensions: 384ft (oa) x 41ft 9in x 19ft (maximum)
Machinery: 2-shaft gas turbines; Olympus TM3B, 56,000hp, Tyne RM1A, 8500hp
Speed: 30kts (maximum), 18kts (cruising)
Armament: 1-4.5in Mk 8 gun; quadruple GWS.24 short-range SAM; 6-12.75in anti-submarine torpedoes (2x3); 4 MM-38 Exocet anti-ship missiles added later
Aircraft: 1 Lynx helicopter
Complement: 175–192

Conclusion

The Type 21 frigate was ultimately a failure in that it did not meet the Royal Navy's requirements for robust ships capable of operating all year round in the North Atlantic. The design was in fact optimised for export, to meet the less demanding needs of customers in the emerging world. This is clear from the propaganda campaign, which focused on the need to boost export sales. Even the difference between better officers' accommodation and lower standards for the lower deck gave the game away. In navies of emerging countries it is generally officers who recommend the purchase of ships, not ordinary sailors.

The design cannot be blamed for the loss of HMS *Ardent* and HMS *Antelope*; they succumbed to punishment which would have sunk any minor warships of similar displacement. Although other ships suffered from the stresses of bad weather, the Type 21s suffered most. There was also strong criticism of their aluminium superstructures, although this was not a major factor in the loss of the two units. But the designers had made such play with the benefits of alloy superstructures that it is hardly surprising that they were held accountable. It turned out that *Amazon* had been badly affected by a fire off Singapore, when alloy ladders melted and

HMS *Ambuscade was attacked by an Exocet during the Falklands War in 1982, but her passive countermeasures succeeded in defeating the attack.*

hampered the firefighters trying to get to the seat of the fire. Much more damaging, however, was the firm evidence of the tendency of alloy panels to be shredded by blast, sending fragments inboard. The fate of the US Navy's large fleet escorts *Worden* and *Belknap* showed just how badly alloy behaved. In the case of the USS *Worden*, while operating off Vietnam she was attacked by a 'friendly' Shrike anti-radiation missile dropped accidentally by an aircraft flying high above her. The missile behaved as it was meant to, homing in on the ship's radiating radars and bursting nearby. The small blast and fragmentation warhead peppered the superstructure, and sent hundreds of alloy fragments into key control spaces. Every system except the machinery was knocked out, and the *Worden* was effectively crippled. The USS *Belknap* was alongside an aircraft carrier in the Mediterranean when she collided with a sponson; a fuel jettison pipe aboard the carrier fractured, drenching the *Belknap* with a large quantity of aviation fuel. The resulting fire virtually melted her entire superstructure, although the steel hull was intact.

The culmination of these accidents and disasters was that alloy for warships went out of favour. After a bitter and largely ill-informed public debate the Royal Navy's designers were vindicated, and the experiment of delegating design-authority totally to a commercial shipyard has never been repeated. As an anonymous senior official in the Ship Department said, 'the builders got out of their depth' when faced with the problems of total ship-design. One of the important differences between commercial designs and official designs is that the commercial designers are using developed weapons whose weight and volume are known. They can usually rely on the customer to make very few changes during the ship's lifetime, and it is also unlikely that she will sustain any battle-damage. In contrast, an official design must conform to the Staff Requirement, which frequently includes new weapons which may turn out heavier or bigger than originally claimed. This was reflected in the much larger and better-armed Type 22 design, and it hardly a coincidence that when four ships—two destroyers and the two Type 21s—were replaced, it was by four even more powerful Type 22s.

The US Navy also abandoned alloy superstructures, and a major point was made for the large *Arleigh Burke* (DDG-51) Aegis destroyers: they were the first ships for many years to be built with all-steel superstructures.

La Combattante type fast-attack craft

Various navies 1968–2002

The Combattante II and Combattante III series of missile-armed fast attack craft (FACs) were in their day regarded as the way ahead for naval warfare. Their combination of high speed and powerful anti-ship missiles would render all frigates, destroyers and major warships obsolete. As we now know, this didn't happen, so it is an opportune moment to ask, what went wrong?

The Soviet Navy pioneered the missile-armed fast patrol boat (FPB), now known as the fast attack craft (FAC), producing the Project 183R design from the late 1950s to the end of 1965. Designed by Y I Yakhunin to meet a requirement approved in August 1957, it was soon noted by NATO as the Komar ('Mosquito') type. On a short 25.5m hull two box-launchers for P-15 anti-ship missiles (known to NATO as 'Styx') were sited aft, while defensive armament comprised a twin 2M-3M 25mm gun mounted on the forecastle. Four M-50F diesels developed 4800 bhp, equivalent to 39 knots. Endurance was a miserable 1000 nautical miles at 12 knots (500 nm at full speed), and they were intended to spend no more than five days at sea.

PG154 of the Federal German Navy, one of its 20 La Combattante IIs.

THE WORLD'S WORST WARSHIPS

The Soviet-built Komar class fast attack craft was the first example of a whole genus of cheap vessels that packed a big punch, but were lacking in many important qualities sought in warships.

NATO and the leading non-aligned navies were somewhat dismissive, seeing the Komar as suitable only for coast defence. In all 58 were built in Leningrad and 52 in Vladivostok, and large numbers were given to 'satellite' and pro-Soviet navies as military aid.

Project 183R might have remained no more than a Cold War curiosity but for a momentous engagement on 21 October 1967, when the elderly Israeli destroyer *Eilat*, patrolling off Port Said, was sunk by missiles fired from a pair of Egyptian Komars in the harbour. The naval world was thrown into a massive panic, without much serious analysis of the engagement, ignoring the *Eilat*'s lack of any modern defences, her rashness in loitering in daylight off a hostile port, and so onetc. Various solutions were adopted; several countries, including Israel, funded a series of anti-ship missiles, notably the French MM-38 Exocet sea-skimmer. The Federal German Navy took the threat particularly seriously, being close to Soviet bases in the Eastern Baltic, and asked the noted specialist Lürssen shipyard in Bremen to design and build a 45-metre missile-armed fast patrol boat (FPB). As a close political ally of France, the West German Government felt confident that it could persuade the French to supply Exocet missiles for the programme.

The French, however, put commercial advantage before Franco-German solidarity, and insisted that a design from the Constructions Mécaniques de Normandie (CMN) design shipyard would be the sine qua non of the deal. This turned out to be just 2 metres longer than the Lürssen design, thus pre-empting any accusation of infringement of design patents or intellectual property rights. To add insult to injury, the French Government insisted that half of the 20 FPBs must be built by CMN. The name Combattante was chosen for the new design, because a small prototype patrol craft of that name had carried out the first sea trials of the Exocet missile. Hence the designation Combattante II, although there was no commonality between the two.

The key to the design was the MM-38 sea-skimming missile. It had been 'on the drawing board' of Aérospatiale's Missiles Division but the French Navy was not interested in funding its development, and potential export customers could not afford to. Then came the sinking of the Eilat, and the immediate reaction of the President of the Missiles Division to his Board of Directors is alleged to have saidbeen to his Board of Directors, 'Gentlemen, a great day has dawned for Aérospatiale. Telephone the bankers and get a loan to fund development'. The result was a weapon travelling at a speed of Mach 0.9, with a maximum range of 22.5 km and armed with a 165 kg fragmentation warhead. 'Styx' was by comparison a crude weapon, very large and heavy, and needing continuous radar guidance from launch to impact. Exocet's great advantage was its inertial navigation system, which made it a 'fire and forget' weapon. Because it weighed only 735 kg the new generation of missile-armed craft could carry four MM-38 Exocets. Full development of Exocet started in August 1968, and four months later large orders were placed by the Royal Navy and the French Navy (300 each), and a smaller order by the Hellenic Navy. Similar missiles were developed for the US Navy (Harpoon), Israel (Gabriel), Italy (Otomat) and Norway (Penguin), all of which also armed Combattantes and their derivatives.

Lürssenwerft appealed to the Federal Government for support in claiming compensation for what it saw as blatant pirating, but the Bonn–Paris Axis was regarded as too valuable to put at risk, and the Bremen yard was told to bite on the bullet. CMN's marketing effort was unaffected by any such worries, and many navies, particularly those small navies looking for a cheap 'equaliser', the traditional Holy Grail of the underdog.

Although variations in weapons and electronics meant that there was no absolute standard, the typical layout was an Oto-Melara (now Otobreda) 76mm L/62 76mm Compact gun, four MM-38 missiles in box-launchers amidships and a Breda (now Otobreda) 40mm L/70 twin mounting. The most favoured propulsion plant was four MTU fast-running diesels driving four shafts. Inevitably rival shipyards cashed in on the concept, creating their own 'Combattante Types', while Lürssen responded with its own TNC 45 design. But CMN had achieved the marketing ideal of forcing every competitor to use its product-name as a generic description.

In all six navies bought a total of 67 Combattante IIs (built 1968–74):

Bundesmarine	20
Hellenic Navy	10
(Imperial) Iranian Navy	12
Israeli Defence Forces	12
Royal Malaysian Navy	4
Libya	9

The limitations of the design led to an expanded design, the 56-metre Combattante III. The longer hull improved seakeeping, and provided more internal volume, although the basic layout remained the same: a 76mm gun forward, missiles amidships and an air defence gun mounting aft. In all 43 were built for six navies (1975–1990):

Hellenic Navy	10
Israeli Defence Forces	15
Tunisian Navy	3
Qatar Emiri Navy	3
South African Navy	9
Nigeria	3

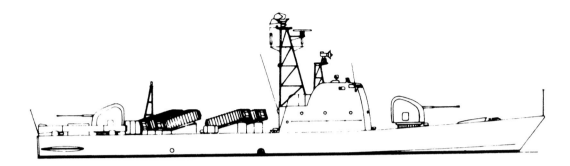

Ipopliarchos Konidis is an example of a La Combattante II bought by the Greek Navy. (Line drawing 1/350 scale.)

COMBATTANTE II FAST ATTACK CRAFT

Combattante II built 1968-74 (Dimensions differ slightly and armaments vary; this is a typical set of figures)

Displacement;:	234 tons (standard), 263.8 tons (full load)
Dimensions:	47 m (oa) x 7.0 m x 2.2 metres
Machinery:	4-shaft 12,000 bhp MTU diesels
Speed:	38.5 knots (maximum in light condition)
Armament:	4 MM-38 Exocet anti-ship missiles (4 x 1), 1-76mm L/62 Oto-Melara Compact gun, 2- 40mm L/70 Breda Compact guns (1 x 2)
Fuel:	33 tons diesel oil
Endurance:	570 nm @ 30 knots; 1600 nm @ 15 knots
Complement:	30

The Combattante III design gained notoriety when five building for Israel were detained at Toulon after the French Government applied a United Nations embargo on the supply of military hardware to areas of conflict. The five FPBs had been built by CMN and were working up under French naval supervision. One night in December 1969, their crews made a daring escape and headed for Haifa, where they were received as heroes. It was assumed at the time that there had been some sort of collusion with French sympathisers, or extremely lax security, and there was talk of a search for the culprits, but the matter was soon forgotten.

The eventual solution was to grant a licence to Israeli Aircraft Industries (IAI) to allow more to be built at the Haifa Shipyard as the *Sa'ar 4* type, better known as the Reshef class. As a violator of UN embargoes, Israel had no hesitation in selling Reshefs to South Africa, sharing construction with a Durban shipyard. The South African Navy's top brass had been debating for some time what type of future surface combatant it wanted, when the Israeli Defence Force transferred some Reshefs from the Mediterranean to the Red Sea, via the Cape of Good Hope. According to hallowed tradition in the South African Navy, the then Defence Minister, P W Botha was watching the evening television news and saw the Israeli missile boats arriving at Simon's Town. To the Navy's horror he told the admirals that he had decided to buy a dozen Reshefs as he was tired of listening to the arguments on the respective merits of corvettes and frigates. Later the Navy managed to stop the programme at nine boats, but in most naval officers' opinion that was nine too many.

Of all the areas of the world unsuited to the Combattante type of warship, the waters around the Cape of Good Hope must be the worst. The seas are steep and even relatively calm weather tests short and lightly constructed hulls.

COMBATTANTE III FAST ATTACK CRAFT

Combattante III built 1975-1990 (Dimensions differ slightly and armaments vary; this is a typical set of figures.)

Displacement:	359 tons (standard), 425 tons (full load)
Dimensions:	56.2m (oa) x 7.9m x 2.5m
Machinery:	4-shaft MTU diesels, 18,000 bhp
Speed:	36.5 knots (maximum, light)
Armament:	4 MM-38 or MM-40 Exocet anti-ship missiles, 1 76mm L/62 Oto-Melara Compact, 2 Breda-40mm Breda L/70guns (1 x 2); 4-x 30mm guns (2 x 2)
Endurance:	700 nm @ 32 knots; 2000 nm @ 15 knots
Complement:	35

CONCLUSION

The missile boat 'revolution' of the 1970s was in reality only the second resurrection of the *poussière navale* ('naval dust') theories associated with the French Navy's Jeune Ecole a century earlier. Then it was the steam torpedo boat which was to sweep the battleship from the seas. Both the Marine Nationale and the Royal Navy built large numbers of torpedo boats in the late 1870s and throughout the 1880s. In service they proved flimsy and unreliable, very susceptible to damage on exercises in anything but a flat calm. They also proved useless for scouting as the view from their low bridges was very restricted. The answer to the torpedo boat 'menace' proved to be the 'torpedo boat destroyer' (TBD), introduced by the Royal Navy in 1893. By doubling the displacement the TBD's seakeeping was improved, and allowed a weight margin for a heavier gun armament. Within a few years the TBD rendered the torpedo boat obsolete; it could destroy hostile torpedo boats by gunfire before they came within torpedo range and then go on to make a torpedo attack themselves on the enemy's fleet.

In the First World War the Royal Navy and the Italian Navy pioneered a new type of petrol-engined torpedo boat, known as Coastal Motor Boats (CMBs) in the Royal Navy and *Motobarca Armata Silurante* (MAS) in Italy. These craft caused great excitement, particularly for young officers, and great things were expected of them. *MAS.15*, commanded by Lieutenant Luigi Rizzo, achieved immortality by sinking the Austro-Hungarian dreadnought Szent Istvan in the Adriatic on 10 June 1918. In 1936 she was preserved and put on permanent show in a museum in Rome. The CMBs' main achievement was a daring raid on the Bolshevik Fleet in Kronstadt on 18 June 1919 during the Intervention War. Although they were handled bravely four 55-ft CMBs were sunk in return for the sinking of the old cruiser *Pamyat Azova*, which was serving as a depot ship.

The brilliant feat of *MAS.15*, and even the meagre results of the CMB raid on Kronstadt, led to a belief that motor torpedo boats could sink large ships with impunity. The German Navy developed its excellent series of Schnellboote (known to the British for some reason as E-boats) but they relied on a combination of moonless nights and calm weather to function successfully. The new generation of MAS-boats achieved remarkably little for the money spent on them, the exploits of the US Navy's PT-boats in the Pacific were hugely over-estimated, and the Royal Navy's Coastal Forces were known as 'Costly Farces' until they learned how to make best use of their MTBs (motor torpedo boats) and MGBs (motor gun boats). What should have been obvious even by 1918—that aircraft were the deadliest threat to MTBs—seemed to be ignored by their supporters. On 11 August 1918 a force of CMBs operating off Terschelling on the Dutch coast was slaughtered by German seaplanes, and air attack finished off the *Schnellboote* in 1944–45.

Post-1945 developments in the Royal Navy focused on improved propulsion (diesels instead of the lethal petrol engines), and more powerful armament, but eventually the Royal Navy wound up its FPB activities in the early 1960s. Since then there have been two engagements between FACs and aircraft to demonstrate that air attack is still lethal. The first was on 24–25 March 1986, when US Navy aircraft launched retaliatory strikes against the Libyan Navy, using air-launched AGM-84 Harpoon missiles to sink a Soviet 'Nanuchka'-type 59-metre corvette and disable a second, and sink a Combattante II and disable another, for no loss to themselves. Even more graphic a demonstration was the so-called 'Battle of Bubiyan Channel,' during the Gulf War in 1991. Virtually the entire Iraqi Navy, a total of 37 ships, was attacked by three Royal Navy Lynx helicopters armed with Sea Skua air-to-surface missiles, leaving them to be finished off by US Navy aircraft armed with Maverick missiles. The Iraqi vessels sunk or disabled by Sea Skuas included five ex-Kuwaiti Navy TNC 45 Lürssen-built FACs captured during the invasion in August 1990, as well as Iraqi Spasilac landing craft and two Zhuk-class patrol boats.

Other drawbacks of FACs are reliability and habitability. To keep a minimum force ready to go to sea, it is necessary to own a relatively large force. The cramped conditions and lively motion in a seaway take their toll on morale and reduce competence in handling complex elec-

The greatest threat to fast attack craft like the La Combattantes comes from the air. In 1986 aircraft of the USN sank one of Libya's La Combattante II in the Gulf of Sidra.

tronics. Small dimensions also make for difficult maintenance, and the light construction gives FACs a relatively short service-life. The cost of upkeep makes nonsense of claims about cheapness, although some operators in emerging countries choose to ignore the problems in the hope that everything will be made to work by negotiating an overhaul contract with the original supplier. A fresh coat of paint and a display of flags can hide a multitude of shortcomings. The Israelis created a very sophisticated support complex for the South Africans to get around the problem of 'low availability' of spare parts, and front-rank navies such as those of Germany and Sweden also spend money on their FACs, but many minor navies care more about prestige than efficiency.

If proof were needed that the small FAC has had its day, navies are turning to larger corvettes with a length in the region of 62 metres. This solves some of the problems of habitability and seakeeping, and is the only way to improve defences against helicopter-attack. It is ironic that the Israelis, who did so much to preach that the days of the frigate and destroyer were numbered, were the first to break ranks by ordering the 86-metre Sa'ar V corvettes in 1989. A number of 'anti-missile missile' systems and gun-based close-in weapon systems (CIWS) are appearing in this new generation of corvettes—regarded as too large only 20 years ago. If there is one lesson to be derived from the story of the steam torpedo boat, the MTB and the FAC, it is that no 'ultimate weapon' lasts very long. The only surprise is that the alleged supremacy of small strike-craft keeps recommending itself to the dissidents and 'Young Turks' in the naval community.

Glossary

barbette	A fixed steel tower rising from the bottom of the hull to the deck), with guns firing over the edge and revolving inside the (usually) cylindrical tower
bhp	brake horsepower (diesel engines)
Calibre	Either the internal diameter of the gun barrel, or a way of indicating the length of the barrel, expressed as multiples of the barrel diameter
CT	conning tower (in surface ships an armoured steering position)
ihp	indicated horsepower (19th-century ships)
kt(s)	knot(s)
ML	muzzle-loading, smooth-bore gun
QF	quick-firer, -firing (a gun with a semi-automatic breech mechanism, firing 'fixed' ammunition
RML	rifled muzzle-loading gun
shp	shaft horsepower (steam turbines
turret	Originally a circular revolving armoured shield resting on rollers, shaped like a pillbox. By the late-19th century it was a 'hooded barbette', an enclosed structure revolving on the top of the barbette (q.v.)

BIBLIOGRAPHY

Alden, Cdr John D, USN, Rtd, *American Steel Navy* (United States Naval Institute Press, 1972)
—— *The Fleet Submarine in the US Navy* (United States Naval Institute Press, 1979)
Allen, Francis J, 'The US Monitors', *Warship VII* (Conway Maritime Press, 1983)
——, 'USS Vesuvius and the Dynamite Gun', *Warship XII* (Conway Maritime Press, 1988)
Brown, David K, 'The Design and Loss of HMS *Captain*', *Warship Technology 7* (London, 1989)
——, (ed.), *The Eclipse of the Big Gun: The Warship 1906–1945*, Conway's History of the Ship series (Conway Maritime Press, 1992)
——, 'The Russo-Japanese War: Technical lessons as perceived by the Royal Navy', *Warship 1996*, (Conway Maritime Press 1996)
——, *Warrior to Dreadnought: Warship Development 1860-1905* (Chatham Publishing, 1997)
Brown, J D, *Carrier Operations in World War II* (Revised edition in 2 vols, Ian Allan 1974)
Brook, Peter, 'Armoured Cruiser versus Armoured Cruiser: Ulsan, 14 August 1904', *Warship 2000-2001* (Conway Maritime Press, 2000)
Burt, R A, *British Battleships of World War One* (Arms and Armour Press, 1986)
Campbell, John, *Naval Weapons of World War Two* (Conway Maritime Press, 1985)
Canney, Donald L, *Lincoln's Navy* (Conway Maritime Press, 1998)
Chesneau, Roger, (ed.), *Conway's All the World's Fighting Ships 1922-1947*, (Conway Maritime Press, 1980)
——, and Eugene M Kolesnik (eds), *Conway's All the World's Fighting Ships 1860-1905* (Conway Maritime Press, 1979)
Chumbley, Stephen, (ed.), *Conway's All the World's Fighting Ships 1947-1995* (Conway Maritime Press, 1995)
Garzke, William H, Jr, Robert O Dulin, Jr, and David K Brown, RCNC, 'The Sinking of the Bismarck: An analysis of the damage', *Warship 1994*, (Conway Maritime Press 1994)
Giorgerini, Giorgio, and Augusto Nani, *Gli Incrociatori Italiani 1861–1975* (Edizioni Ufficio Storico della Marina Militare, 1976)
Gray, Randal, (ed.), *Conway's All the World's Fighting Ships 1906-1921* (Conway Maritime Press, 1985)
Friedman, Norman, *Modern Warship Development and Design* (Conway Maritime Press, 1979)
——, (ed.), *Navies in the Nuclear Age: Warships since 1945* Conway's History of the Ship series (Conway Maritime Press, 1992)
Jordan, John, 'The French Cruiser *Algérie*', *Warship 2001-2002* (Conway Maritime Press, 2001)
Jung, D, and H Jentschura, (J D Brown and A Preston, trans.) *Ships of the Imperial Japanese Navy* (Arms & Armour Press, 1977)
Jurens, William J, 'The Loss of HMS *Hood*: A re-examination', *Warship International* XXIV/2 (1987)

Lambert, Andrew, (ed.), *Steam, Steel and Shellfire: The Steam Warship 1815–1905* Conway's History of the Ship series (Conway Maritime Press, 1992)

Layman, R D and Stephen McLaughlin, *The Hybrid Warship*, (Conway Maritime Press, 1990)

Lipiett, John, *Type 21*, (Ian Allan, 1990)

McBride, Keith, 'The Weird Sisters', *Warship 1990*, (Conway Maritime Press, 1990)

McLaughlin, Stephen, 'From Riurik to Riurik: Russia's Armoured Cruisers', *Warship 1999-2000* (Conway Maritime Press, 1999)

Marriott, Leo, *Royal Navy Frigates Since 1945* (Ian Allan, 1990)

Preston, Antony, 'Soviet and Russian Air-Independent Submarines', *Warship 2000-2001* (Conway Maritime Press, 2000)

Raven, Alan 'Aspects of Japanese Warship Design', *Warship I* (Conway Maritime Press, 1977)

Roberts, J A, *Battlecruisers* (Chatham Publishing, 1997)

Sieche, Edwin, 'The Walter Submarine-1' *Warship V*, (Conway Maritime Press, 1981)

US Naval History Division *Monitors of the U.S. Navy, 1861-1937* (prepared by Richard H. Webber) (Government Printing Office, 1969)

Watts, A J, and B G Gordon, *The Imperial Japanese Navy* (Macdonald, 1971)

Wilson, Michael, The Walter Submarine-2', *Warship V*, (Conway Maritime Press, 1981)

INDEX

Figures in *italics* refer to illustrations and those in **bold** refer to diagrams.

AA class fleet submarines 73-7, *74*
   specifications 74-5
AA-1 (US submarine) 73
   specifications 74-5
AA-2 (US submarine) 73
   specifications 74-5
AA-3 (US submarine) 73
   specifications 74-5
Achilles, HMNZS 121
Active, HMS (corvette) 44
Active, HMS (Type 21 frigate) 172, 173
   specifications 174-5
Admiral Baudin 67
Admiral Graf Spee 118, *119*, *120*, 120-1, *122*
   specifications 122
Admiral Hipper 122, 151
Admiral Scheer **121**
   specifications 121-2
Afridi, HMS 57
air power 156-7
aircraft 103, 104, 108, 109, 110, 111, 113, 125, 150, 153, 155
aircraft carriers
   aircraft capacity 132, 132-3, 154, 155, 158
   armament 131, 133, 153
   armour 153-4
   doctrine 111
   flight decks 131, 133, 153-4, 158
   French navy 70-1, 113
   hangers 153-4, 155, 158
   Imperial Japanese Navy 111, 130-5, *131*, *132*, *133*, **134**, *145*, 145-6, 153, 154
   Imperial Russian Navy **134**
   Royal Navy *92*, 93, *93*, 111, 132, 150, 153-8, *154*, **156**, *157*
   US Navy 72, 111, 135, 153, 154, 158
Ajax, HMS 121
Akagi 131-2, 135
Alacrity, HMS 172, 173, **174**
   specifications 174-5
Alberico da Barbiano 125, 125-6, 129
   specifications 126-7
Alberto di Giussano 125, 126, 129
   specifications 126-7
Aleksandrovich, Grand Duke, Aleksei 36
Alexandria, Bombardment of 24
Algérie 116
Allen, USS 80
'Alpha' class nuclear attack submarines 165-70, *169*
   specifications 168-9
Altmark 120
Alwyn, USS 80
Amagi 131-2
Amazon, HMS 172, *173*, 175-6
   specifications 174-5

Ambuscade, HMS 172, 174, *175*
   specifications 174-5
American Civil War 16-20
Ammen, Rear Admiral Daniel 33
ammunition stowage 47
Amphitrite class monitors 20
Anchar class nuclear submarines 167
Annual Manoeuvres, 1885 32
Antelope, HMS 172, 173-4, *175*
   specifications 174-5
anti-submarine warfare 80-1, 171-6
Aoba 137
Archerfish, USS 145
Ardent, HMS 172, 173, *175*
   specifications 174-5
Argonaut, USS 76
Ark Royal, HMS 150, 154
Arkansas class monitors 20
Arleigh Burke, USS 176
armament 12, 32
   18in guns 93, 95
   46cm guns 143-4, *144*
   aircraft carriers 131, 133, 153
   anti-aircraft 59, 99, 113, 114, 125, 152
   anti-submarine 165
   battlecruisers 92, 93
   battleships 53, 67, 68, 99, 100, 148, 152
   cruisers 46, 47, 48, 49, 103, 103-4, 106, 108, 108-9, 110, 112, 113, 114, 137, *138*, 139
   Dahlgren guns 16, 17
   destroyers 57, 59, 60, 79-80
   discharge effects 41
   dreadnoughts 62, 63, 66
   dynamite guns 40-1, 42-3
   fast attack craft 177, 179
   frigates 172
   light cruisers 125
   Parrot rifled guns 19
   Rurik 36, 37
   strategic missiles 165
   submarines 73, 84, 85, 165
Armando Diaz 127
   specifications 127
armour 49
   aircraft carriers 153-4
   anti-submarine 92
   anti-torpedo 62
   battleships 50, 51, 53-4, 96, 97, 143, 151
   cruisers 103, 104, 108, 112, 137
   Krupp's cemented (KC) armour belt 51
   Novgorod 26
   Rurik 36, 37
   USS *Monitor* 16
   Vitse Admiral Popov 26-7
armoured rams 30-4
Armstrong 57
Arrow, HMS 172, 173

   specifications 174-5
Atlanta, USS 19
Austria 61-2
Austro-Hungarian Navy 30, 61-6
Avenger, HMS 172, 173, 174
   specifications 174-5
aviation 70-1, 93, 111, 131
Azuma 38

B class submarines 75-7
Balch, USS 80
Baltic Project 90-1, 95
Baltic Works 36, 37
Barnaby, Nathaniel 31, 44
Barracuda, USS 75
Bartolomeo Colleoni *124*, 125, *126*
   specifications 126-7
Bass, USS 75
Bataan, USS 145
Bath Iron Works 81, 83
battlecruisers 102
   armament 92, 93
   German navy 96, 118
   Imperial Japanese Navy 131
   Royal Navy 90-5, *91*, 119
   US Navy 102
battleship-carriers 111
battleships
   armament 53, 67, 68, 99, 100, 148, 152
   armour 50, 51, 53-4, 96, 97, 143, 151
   French Navy 67, 68
   German navy 11, 97, 99-100, 148-52, *149*, **151**
   Imperial Japanese Navy 38, 52, 142-7, *143*, *144*, *145*, **146**
   Imperial Russian Navy 50-5, *52*, *54*
   Italian Navy 69
   Royal Navy 64, 96-101, *97*, **97**, 148, 149, 150, 152
   US Navy 14, 102, 152
Bayern 148
Béarn 68, 69, 70-1, *71*, 72
   specifications 70, 71-2
Belknap, USS 176
Belleau Wood, USS 145
Benham, USS 80
Bennington, USS 145
Beria, Lavrenti 165
Beverley, HMS 83
Bismarck 11, 99-100, 148-52, *149*, **151**, 155
   specifications 150-1
Bismarck class battleships 148-52, *149*, **151**
Bizerte 69
Black Prince, HMS 12
Black Sea 27-8
Blonde, HMS 86
Bogatyr 37, 38
boilers 46, 49, 58
Bonita, USS 75, 77
Borodino 51, 52, 54, 55

specifications 52-3
Borodino class battleships 50-5, *52*
   specifications 52-3
Botha, P W 180
Branch, USS 83
Brest 69
Bretagne class battleships 67, 68
Broke, HMS 59
Brooklyn, USS 46
Brown, David K 14, 25, 158
Brüning, Chancellor 118
Bubiyan Channel, Battle of 182
Buchanan, USS 81
Bunker Hill, USS 145
Burgoyne, Captain 23-4

Cabot, USS 145
Cachalot, USS 77
Caio Duilio 155
Caldwell, USS 80
Caldwell class destroyers 80-1
   specifications 82
Camanche, USS 18
Campbeltown, HMS 81
Cannery, Donald J 18
Canonicus class monitors 19
Canopus, HMS 11
Captain, HMS 13, 21-5, *22*
   specifications 24
Carraciolo class battleships 69
carrier-cruisers 111
Casco class monitors 19
Cassin USS 80
Cassin class destroyers 79, 80
Catskill, USS 20
central battery ships 30
'Charlie' class nuclear submarines 167
Charleston, attack on 19
Chesapeake, USS 11
Chester class cruiser 102
Cheyenne, USS 20
Chikhachev, Admiral N M 37
Childers, Hugh 24
Chilean Navy 9
China 130
Churchill, Winston 34, 90, 91
Cincinnati USS 107
   specifications 106-7
circular ships 26-9
Clemson class destroyers 81, 83
   specifications 82-3
coastal motor boats 181
Cobra, HMS 57
'Cod War' 172
Cold War, the 9-10
Coles, Captain Cowper Phipps 18, 21-3, 25
Combattante Type Fast Attack craft *177*, 177-83, **180**, *182*
   specifications 180
commerce raiding 35-6, 39, 119, 122
comparisons 9-10
Comus class corvettes 45
Concord, USS 107
   specifications 106-7
Condottieri class light cruisers *123*, 123-9, *124*, *126*, **128**

specifications 126-7
Confederate States of America 16
Congress, USS 17
Conner, USS 80
Constructions Mécaniques de Normandie 178
Conte di Cavour 155
convoys 36
Conyngham, USS 80
Coral Sea, Battle of the 144
corvettes 183
costs 10, 12, 26, 57, 58, 134-5, 169-70, 171
Courageous, HMS 92, *92*, 93, 94, 95
   specifications 94
Cramp shipbuilders 78, 80, 83
Craven, USS 80
Crimean War. *see* Russian War of 1854-56
cruisers 9, 11
   armament 46, 47, 48, 49, 103, 103-4, 106, 108, 108-9, 110, 112, 113, 114, 137, *138*, 139
   armour 103, 104, 108, 112, 137
   French navy 112-16, *113*, **115**
   Imperial Japanese Navy 11, 136-41, 137, *138*, **140**
   Imperial Russian Navy 11, *35*, 35-9, **39**, 45, 48, 49
   Royal Navy 44-9, *45*, *46*, **48**, 106
   Royal Swedish Navy 108-11
   US Navy 102-7, *105*, *106*
cruizing 44
Cumberland, USS 17, 30
Cummings, USS 80
Cunningham, Admiral Andrew Browne 155
Cushing, USS 80
Cuttlefish, USS 77

Dahlgren, Admiral John A 17
Daniels, Josephus 104
Dante Alighieri 62, 63
Danton class battleships 68
Danubius shipyard 63, 64-5
Dardanelles Campaign 92
de Bon, Admiral 69
de Lôme, Dupuy 12
Decatur, Stephen 11
'Delta I' class nuclear submarines 167
design responsibility 13, 23, 25, 176
design teams 14
destroyers
   armament 57, 59, 60, 79-80
   French navy 124
   German navy 59
   Royal Navy 11, 14, 56-60, 80, 81, *81*, 83
   US Navy 78-83, *79*, *81*, *82*, 102
Detroit, USS 107
   specifications 106-7
Deutschland 118, 118-19, 120
   specifications 121-2
Deutschland class 'pocket battleships' 117-22, *119*, *120*, **121**
   specifications 121-2
d'Eyncourt, Sir Eustace Tennyson 91, 96
Diadem class protected cruisers 49
Dictator, USS 19

Dnieper River 26, 27
Dolphin, USS 76
Dorsetshire, HMS 152
double-turreted monitors 19
Dover
   action off, April 1917 59
Dreadnought, HMS 13-14, 32, 56
dreadnoughts 13-14, 32, 56, *61*, 61-6, 67-72, **70**, 181
Druzni 81
Dubasov, Admiral 37
Dumaresq, Captain 58
Duncan, USS 80
Dunn, W J 31
Dunning, Squadron Commander 93
Duquesne 112-13, *113*, 113-14
   specifications 114-15
Duquesne class heavy cruisers 112-16, *113*, **115**
   specifications 114-15
dynamite 40, 42
dynamite cruisers 40-3

Eagle, HMS 71, 155
Earle, Admiral 103
Eastern Solomons, Battle of the 134
'Echo' class nuclear submarines 165
Edgar class cruisers 48
Egyptian navy 178
Eilat 178, 179
Elder, John 29
Elli (former *Eugenio di Savoia*) 128
Emanuele Filiberto Duca d'Aosta 127-8
Emden 117
Enterprise, USS 145
Ericsson, John 16
Ericsson, USS 80
Ersatz Hannover 148
Ersatz Preussen (later *Deutschland*) 118
Ersatz Schleswig Holstein 148
Ersatz Yorck class battlecruisers 96
Ersh class nuclear submarines 167
Erzherzog Franz Ferdinand class pre-dreadnoughts 62
escort carriers 135
Esmeralda 9
Essex, USS 145
Eugenio di Savoia 128
Evans, Commander Edward 59
Excalibur, HMS 161, *161*, 163
Exeter, HMS 121, 122
Exocet missiles 173, 174, 179
Explorer, HMS 161, 163
Explorer class submarines 161, *161*, 163
   specifications 161

Falklands, Battle of the 90
Falklands War, 1982 9, 172, 172-4
fast attack craft *177*, 177-83, **180**, *182*
fast patrol boats 178, 182
Fearless, HMS 86
Ferdinand Maximilian 30
fighter direction 110
fire control 53, 94, 110, 152
Fisgard III, HMS (former HMS *Terrible*) 47
Fisher, Admiral Sir John 13-14, 36, *56*, 56-

7, 58, 60, 84, 90-1, 92, 94-5
Flandre 68
    specifications 70
Fletcher, Admiral Frank 104
floatplanes 103, 104, 109, 110, 111, 113, 125
Flying Scud (later HMS *Swift*) 57
Ford River Shipbuilding 78
Forges et Chantiers de la Méditerranée 71
Formidable (French battleship) 67
Formidable, HMS (British aircraft carrier) 154, 155, 157, 158
Fourth Fleet Incident, the 137-8
France 12, 112, 115-16
Franklin, USS 145
French navy 12, 36, 65, 99
    aircraft carriers 70-1, 113
    cruisers 112-16, *113*, **115**
    destroyers 124
    dreadnoughts 67-72
    fast attack craft 178-9
    men o'war 10-11
    submarines 87
    torpedo boats 181
Friedman, Norman 10
Furious, HMS *93*, 93, 94, 95
Furutaka 137
Furutaka class cruisers 11, 137
Fylgia 108

G.42 (German destroyer) 59
G.85 (German destroyer) 59
Gascogne 68
    specifications 70
General Board, the 79-80, 102
German navy 66
    battlecruisers 96, 118
    battleships 11, 97, 99-100, 148-52, *149*, **151**
    destroyers 59
    fast attack craft *177*
    fast patrol boats 178
    High Sea Fleet 12, 64, 119, 148
    'pocket battleships' 117-22, *119*, *120*
    rearms 117-20
    Schnellboote 181
    submarines 87, 159-60, 163
    and the Treaty of Versailles 117
Giovanni delle Bande Nere 125, 126
    specifications 126-7
Giuseppe Garibaldi 128
Glorious, HMS *91*, 92, 93, 94, 95
    specifications 94
Gneisenau 93, 118
Goodall, Sir Stanley 100, 122
Gorshkov, Admiral Sergei 10, 12, 167, 170
Gotland, HSwMS 108-11, *109*, 149
    specifications 109
Grazhdanin (former *Tsessarevitch*) 50
Great Britain 12
    and Japan 130, 136, 142
Greek navy 128, 179, **180**
Greer, USS 83
Gromoboi 37, 38, 39
gunnery 53, 94-5
Gwin, USS 80

Hampton Roads, Battle of 17-18, 20, 30
Hannah Donald, shipbuilders 14
Harwood, Commodore Henry 121
Hatsuyuki 138
Heligoland Bight, action in the 92-3
Henderson, Rear Admiral Reginald 153, 154
Henri IV 62
Hermes, HMS 111
Hindenburg 97
Hindenburg, Paul von 118
Hiryu 135
Hiryu class aircraft carriers 133, 135
Hitachi Maru 37
Hitler, Adolf 118, 119, 120, 148
Hood, HMS 64, 96-101, *97*, **97**, 148, 149, 150
    specifications 97
Hood, Rear Admiral Horace 98
Hornby, Admiral Geoffrey Phipps 32
Hornet, USS 145
Horthy, Admiral 61
Hosho 131, *131*, 132
'Hotel' class nuclear submarines 165, 170
Houston, USS 139
hull design 9, 26, *27*, 28-9, 33, 50, 56, 135, 138-9, 140
    submarines 73, 75
Hull No. 111 143
Hull No. 797 143
Hungary 61

Idzumo 38
Iessen, Admiral 38
Illustrious, HMS 154, 155, 157, 158
Illustrious class aircraft carriers 154
Imperator Alexander III 51, 52, 55
    specifications 52-3
Imperial Japanese Navy 13, 66, 81, 112, 115, 119
    aircraft carriers 111, 130-5, *131*, *132*, *133*, **134**, *145*, 145-6, 153, 154
    battlecruisers 131
    battleships 38, 52, 142-7, *143*, *144*, *145*, **146**
    cruisers 11, 136-41, *137*, *138*, **140**
    at Tsushima 52, 53, 130
    at Ulsan 38, 39
Imperial Russian Navy 35-9. *see also* Soviet Navy
    Baltic Fleet 50, 51, 136
    battleships 50-5, *52*, *54*
    coast defence ships 26-9
    cruisers 11, *35*, 35-9, *39*, 45, 48, 49
    submarines 159
    at Tsushima 51-2, 53-4, 130, 136
    at Ulsan 38, 39
Implacable, HMS 154, 155, 156
    specifications 156
Implacable class fleet aircraft carriers 153-8, *154*, **156**, *157*
    specifications 156
Impregnable II, HMS (Former HMS *Powerful*) 47
Inconstant, HMS 24, 44
Indefatigable, HMS 154, *154*, 155, 157, 158

    specifications 156
Indochina 71, 114
Indomitable, HMS 155, 156, 157, 158
industrial capability 12-14, 72, 146
Inflexible, HMS 86
intelligence assessments 11, 48-9
Intervention War, the 111, 181
Intrepid, USS 145
Invincible, HMS 97, 98
Iowa class battleships 14
Ipopliarchos Konidis **180**
Iraqi Navy 182
Iris, HMS 44
Isla dos Estados 173
Israeli Defence Force 180, 183
Italian navy 30, 62, 64, 65, 69, 112, 155
    light cruisers *123*, 123-9, *124*, *126*, **128**
    MAS boats 181
    motor torpedo boats 64
Iwami (former *Orel*) 52

J class submarines 84
Jacob Jones, USS *79*, 80
Japan, and Britain 130, 136, 142
Jellicoe, Admiral Sir John 84, 91, 96
John Brown shipbuilders 57
Jutland, Battle of 64, 88, 96, 98, 100, 102

K.1 (British submarine) 86, 88
K.3 (British submarine) *85*, 85
K-3 (Soviet submarine) 165
K.4 (British submarine) 85, 86
K.5 (British submarine) 86, 88
K-5 (Soviet submarine) 165
K.6 (British submarine) 86
K.8 (British submarine) 86
K-8 (Soviet submarine) 165
K-11 (Soviet submarine) 165
K.12 (British submarine) 85
K.13 (British submarine) 86, 88
K.14 (British submarine) 86
K.17 (British submarine) 84, 86
K.22 (British submarine) 86
K.26 (British submarine) 86, *88*
K-123 (Soviet submarine) 168
    specifications 168-9
K-278 (Soviet submarine) 168
K-316 (Soviet submarine), specifications 168-9
K-373 (Soviet submarine), specifications 168-9
K-377 (Soviet submarine) 168
    specifications 168-9
K-423 (Soviet submarine), specifications 168-9
K-463 (Soviet submarine), specifications 168-9
K-493 (Soviet submarine), specifications 168-9
K class submarines 75, 84-9, *85*, 88
    specifications 86-7
Kaga 131, 132, *132*, 135
Kako 137
Kalamazoo class monitors 19
Kamimura, Vice Admiral 38, 39
Kandahar, HMS 126

Katahdin, USS 33, 33-4, 43
    specifications 34
Kerch (former *Emanuele Filiberto Duca d'Aosta*) 127-8
Kerch Straits 26
Kiaou Chou 130
Kiev (renamed *Vitse Admiral Popov*) 26
King George V, HMS 97, 150
Kinugasa 137
Kinugasa class cruisers 137
Klas Fleming 108
Kniaz Suvorov 51, 52, 53-4
    specifications 52-3
Komar type fast patrol boats 177-8, *178*
Kongo class battlecruisers 119
König 52
Kronprinz 50
Kronstadt, raid on 181
Kumano 137, 139, **140**
    specifications 139-40
Kursk 164

la Gloire 12
La Seyne yard 50, 71
Lady Nancy gun raft 21
Ladysmith, Relief of 47
Laird Brothers 23, 25
Landevennec 70
Langsdorff, Captain 120-1
Languedoc 68
    specifications 70
Leander class frigates 171, 172
Lehigh, USS 19, 20
Leninskiy Komsomol 165
Lexington, USS 72, 97
Leyte Gulf, Battle of 139, 145
Libyan Navy 182
light cruisers, Italian navy 123, 123-9, *124*, *126*, **128**
Lincoln, HMS 81
Lion class battleships 152
Lira 167
Lissa, Battle of 30, 62
Littorio 155
Livadia 29, *29*
London Naval Treaties 74, 137, 139, 142
Long, John D 42
Lorient 69
Louis of Battenburg, Prince 90
Luigi Cadorna 127
    specifications 127
Luigi di Savoia Duca degli Abruzzi 128
Lürssenwerft 178, 179
Lütjens, Admiral Guenther 152
Lützow 120, 122

M-92 (Soviet submarine) 161
M-255 (Soviet submarine) 162
M-256 (Soviet submarine) 162
M-257 (Soviet submarine) 162
M-259 (Soviet submarine) 162
M-351 (Soviet submarine) 162
M-352 (Soviet submarine) 162
M-361 (Soviet submarine) 162
M-401 (Soviet submarine) 161, 162
McDougal, USS 80

McKee, Andrew I 76
McKee, Captain Logan 163
magazine explosions 54, 100
Mahan, Admiral 122
Makin Islands, raid on 76
Malta class aircraft carriers 158
Manley, USS 80
Marblehead, USS 106, 107
    specifications 107
Mare Island Navy Yard 76, 81
marinised aluminium alloy 171, 174, 175-6
Martinique 71
MAS-15 64, 181
May, James 24
'May Island, Battle of' 86, 88
Maya 139
Mayrant, USS 78
Memphis, USS 107
    specifications 107
Mercury, HMS 44
Merrimack, USS 17-18
Mers-el-Kebir 99, 116
Meteorite, HMS (former *U-1407*) 160, 160-1, 163
Miantonomoh, USS 20
Miantonomoh class monitors 19, 20
Midway, Battle of 135, 139, 144, 145
Mikuma 137, 138, *138*, 139
    specifications 139-40
Milne, Sir Archibald 23
Milwaukee, USS 106, 107
    specifications 106-7
Milwaukee class monitors 19
mines and minelaying 108
MM-38 missile 179
Mobile Bay, Battle of 19
Mogami 137, 138, 139
    specifications 139-40
Mogami class cruisers 136-41, *137*, **140**
    specifications 139-40
Monadnock, USS 20
Monaghan, USS 78
Monarch, HMS 22, 23, 24
'monitor fever' 18, 20
Monitor, USS 16-18, *17*, 20, 21
    specifications 18
monitors 16-20, *17*, 34
Montecuccoli, Admiral Count Rudolf 62, 63
Monterey, USS 20
Morskoi Techniceskii Komitet (MTK) Naval Technical Committee 36, 37
Motobarca Armata Silurante (MAS) torpedo boats 181
Murena class nuclear submarines 167
Murmansk (former USS *Milwaukee*) 106
Musashi 144, 145, 147
    specifications 146
Mussolini, Benito 123-4, 128
Muzio Attendolo 127

Naniwa 38
Napoleonic Wars 10-11, 12, 35-6
Narwhal, USS 76
Nautilus, USS (fleet submarine) 76
Nautilus, USS (nuclear submarine) 87, 165

naval architects 9, 11
Naval War College 102-3
Navigatori class scout cruisers 124
Nelson, HMS 97, 99
Neptune, HMS 126
Nevada, USS 97
Nicholson, USS 80
Nikolaev naval shipyard 27
Nile, Battle of the 11
Norfolk, HMS 149, 150
Norfolk Navy Yard, Virginia 16
Normandie 68
    specifications 70
Normandie class dreadnoughts 67-72, **70**
    specifications 70
Normandy landings 114
'November' class nuclear submarines 165, *166*
Novgorod 26-9
    specifications 28
Novik 38

O'Brien, USS 80
O'Brien class destroyers 80
Okinawa 145
Omaha, USS 107
    specifications 106-7
Omaha class scout cruisers 102-7, *105*, *106*
    specifications 106-7
Onondaga, USS 19
Operation 'Judgement' 155
Operation Rheinübung 149-50
operational environment 14-15
Orel 51, 52, 53, 54, *54*
    specifications 52-3
Otranto Barrage, raids on the 64

Palestro 30
Pamiat Azova 36, 181
panzerschiffe (German armoured ships) 117, 118, 122
'Papa' class nuclear submarines 167
Paris, Treaty of, 1856 35
Parker, USS 80
Parkington, Sir John 23
Passaic class monitors 18-19, 20
    specifications 19
Patapsco, USS 19
Patterson, USS 78
Paulding group destroyers 78
Pearl Harbor, attack on 144
Peck, Commander Ambrose 59
perceived threats 10-11, 48
Peregodov, V N 167
Perkins, USS 78
Perth, HMAS 139
Petrov, A B 167
Pitzinger, Fritz 64, 65
Plavnik class nuclear submarine 168
Pneumatic Dynamite Gun Company, the 40, 41
'pocket battleships' 117-22, *119*, *120*
Podkapelski, Captain Janko Vukovic de 65
Polyphemus, HMS *31*, 31-3, 34, 43
    specifications 33
Pope, USS 134

Popov, Vice Admiral A A 26, 29
Popper, Siegfried 62, 64, 65
Port Arthur, siege of 37-8, 50
Porter, USS 80
Portsmouth Dockyard 14
Portsmouth Navy Yard, USA 75
Powerful, HMS *45*, *46*, 46, 47, 49
    specifications 47-8
Powerful class protected cruisers 44-9, *45*, *46*, *48*
    specifications 47-8
Premuda Island 64
Prince Albert, HMS 18, 21
Prince of Wales, HMS 99, 149, 150, 152
Principal Naval Overseer (PNO) 13
Prinz Eugen (Austro-Hungarian dreadnought) 65
    specifications 65-6
Prinz Eugen (German heavy cruiser) 99-100, 149-50
privateering 35
prizes 10, 11
Project 183R 178
Project 612 submarines 161
Project 615 submarines 161-2
Project 616 submarines 162
Project 617 submarines 162
    specifications 162
Project 627 submarines 165, *166*
Project 637 submarines 162
Project 658 submarines 165, 170
Project 659 submarines 165
Project 661 submarines 167
Project 667 submarines 165, 167
Project 667B submarines 167
Project 670 submarines 167
Project 671 submarines 167
Project 685 submarines 168
Project 705 submarines 167-8, 169-70
Puritan, USS 19, 20

'Quebec' class submarines 161-2
Queen Elizabeth class battleships 96, 99

radar 114
Raeder, Vice Admiral Erich 118, 148
Raimondo Montecuccoli 127
Raleigh, HMS (British frigate) 44, 106
Raleigh, USS (US scout cruiser) *105*, 106, 107
    specifications 106-7
Re d'Italia 30
Reagan, President Ronald 170
Reed, Sir Edward 21, 22, 23, 24, 25, 26
Renown, HMS 90, 99
Repulse, HMS 90, 98, 99
Reshef class fast attack craft 180
Richmond, USS 107
    specifications 106-7
Riga, Gulf of 52
River class destroyers 56, 60
River Plate, Battle of the *120*, 120-1, 122
Rizzo, Lieutenant 64
Roanoake, USS 19
Robinson, Sir Spencer 22, 23, 24, 25
Rodionov, N E 36

Rodney, HMS 97, 99, 150
Roe, USS 78
Ronarc'h, Admiral 69-70
Rossia 37, 38, 39
Rota, General Giuseppe 128
Rover, HMS 31
Royal Air Force 155, 156
Royal Corps of Naval Constructors (RCNC) 9
Royal Naval Air Service 154-5
Royal Navy 66
    aircraft 155
    aircraft carriers *92*, 93, *93*, 111, 132, 150, 153-8, *154*, *156*, 157
    anti-submarine frigates 171-6, *173*, *174*, *175*
    armoured rams 31-3
    aviation 70, 93
    battlecruisers 90-5, *91*, 119
    battleships 64, 96-101, *97*, 97, *98*, 148, 149, 150, 152
    and the *Bismarck* 99-100, 148, 149-50
    carrier-cruisers 111
    Coastal Forces 181-2
    cruisers 44-9, *45*, *46*, *48*, 106
    design requirements 9
    destroyers 11, 14, 56-60, 80, *81*, *81*, 83
    escort carriers 135
    fast attack craft 179
    fast patrol boats 182
    Fleet Air Arm 155
    monitors 20, 93
    role 44
    and the Royal Swedish Navy 110
    Ships' logs 13
    submarines 75, 84-9, *85*, *88*, 160, *160*, 161, *161*, 163, *164*
    torpedo boats 181
    torpedoes 164
    turret ships 18, 21-5
Royal Sovereign, HMS 18, 21
Royal Swedish Navy 108-11, 149
    submarines 163
    torpedoes 164
Rurik 11, *35*, 35-9, *39*, 45, 48, 49
    specifications 38-9
Rusanov, M G 165, 167
Russian War of 1854-56 21, 26, 110
Russo-Japanese War 37-8, 39, 50, 51-2, 130, 136
    lessons of 53-5
Ryujo 130-5, *133*, *134*
    specifications 134

S-99 (Soviet submarine) 162
sailing ships 12-13
San Jacinto, USS 145
Santiago, Battle of 34
Saratoga, USS 72, 134
Sartorius, Admiral 31
Scharnhorst 93, 118
Scharnhorst class battleships 151
Schley, USS 73
sea trials 14
seasickness, effect of 15
Sevastopol, siege of 28

Shah, HMS 44
Shannon, HMS 11
Shaw, USS 80
Sheffield, HMS 173
Shestakov, Admiral I A 36, 37
Shinano 143, *145*, 145-6, 147
    specifications 146
Shokaku 131
Shokaku class aircraft carriers 133
Sibuyan Sea, Battle of the 139, 145
Sidon, HMS *164*
Singapore 93, 134, 139
Skat class nuclear submarines 167
Skate, USS 145
Slava 52
    specifications 52-3
Smith class destroyers 78, 79
Soryu 135
South Africa 180
South Dakota class battleships 152
Soviet Navy 9-10, 12, 37, 106, 127-8. *see also* Imperial Russian Navy
    fast patrol boats 177-8, *178*
    submarines 161-2, 165-70, *166*, *169*
    torpedoes 164
Spanish-American War 20, 33-4, 42
Spaun, Admiral von 62
speed-margins 53
St Chamond Company 67, 68
St Nazaire Raid 81
Stabilimento Technico Triestino 63
Stalingrad (former *Emanuele Filiberto Duca d'Aosta*) 127-8
Sterett, USS 78
Stimers, Alan 19
Stockton, USS 80
submarines
    air-independent propulsion 163-4
    armament 73, 84, 85, 165
    ballast tanks 73-4
    crush depth 85, 88
    French navy 87
    German navy 87, 159-60, 163
    hull design 73, 75
    hydrogen peroxide 159-64, *160*
    Imperial Russian Navy 159
    nuclear 162, 163, 165-70, *166*, *169*
    Royal Navy 75, 84-9, *85*, *88*, 160, *160*, 161, *161*, 163, *164*
    Royal Swedish Navy 163
    Soviet Navy 161-2, 165-70, *166*, *169*
    US Navy 73-7, *74*, 77, 87, 163, 165
Suffolk, HMS 80, 99, 149, 150
Suffren class heavy cruisers 116
Suzuya 137, 138, 139
    specifications 139-40
Swift, HMS 56-60, *59*
    specifications 60
Sydney, HMAS 125
Szent István *61*, 63, 64, 181
    specifications 65-6

T-1 (US submarine, former *AA-1*) 74
T-2 (US submarine, former *AA-2*) 74
T-3 (US submarine, former *AA-3*) 74
tactics 75, 78, 88

ramming 30, 34
Taganrog, bombardment of 21
Takachiho 38
Taranto 64, 155
Taylor, Rear Admiral David W 104, 106
technical capability 12-14, 42-3, 60
Tecumseh, USS 19
Tegetthof, Admiral 30, 62
Tegetthoff 65
    specifications 65-6
Terrible, HMS 45, 47, 49
    specifications 47-8
Terry, USS 78
Thornycroft 14
threat perception 10-11, 48
Tillman, USS 81
Tirpitz 148, 149
    specifications 150-1
Tirpitz, Admiral Alfred von 12
Tomodzuru (Japanese torpedo boat) 133, 138
Tondern raid, the 93
Tone class cruisers 139
torpedo boat destroyers 181
torpedo boats 31-2, 181
torpedo-cruisers 34
torpedoes 31, 32, 57, 73, 79-80, 84, 112, 164, 165
Toulon 69, 71, 116
Tourville 112-13, 114
    specifications 114-15
Town class destroyers 81, 83
Trenton, USS 107
    specifications 107
Tribal class destroyers 60
Tsessarevitch 50, 54, 62
Tsushima, Battle of 51-2, 53, 53-5, 130, 136, 146
Tucker, USS 80
Tucker class destroyers 80
Tunny, USS 145
turret rams 30
turrets 18
    Cowper Coles 18, 21-2
    USS *Monitor* 16, 17
Type 21 Anti-submarine frigates 171-6, *173*, **174**, *175*
    specifications 174-5
Type XVII U-boats 160, *160*, 163
    specifications 160
Type XVIII U-boats 160
Type XXI U-boats 163
Type XXIV U-boats 160
Type XXVI U-boats 160
U-792 (German submarine) 159
U-793 (German submarine) 159
U-1406 (German submarine) 160, 163
U-1407 (German submarine) *160*, 160-1, 163
Ulsan, Battle of 38, 39
Unbroken, HMS 127
United States of America, and Japan 136
Upright, HMS 127

Urge, HMS 126
US Navy 10, 11, 12, 66, 141, 182
    aircraft carriers 72, 111, 135, 153, 154, 158
    armoured rams 33-4
    battlecruisers 102
    battleships 14, 102, 152
    cruisers 102-7, *105*, *106*
    destroyers 78-83, *79*, *81*, **82**, 102
    dynamite cruisers 40-3
    escort carriers 135
    fleet escorts 176
    monitors 16-20, *17*, 34
    PT-boats 181
    submarines 73-7, *74*, *77*, 87, 163, 165
    torpedoes 164
Ustinov, Dimitri 168

V-1 class submarines 75-6
V-4 (US submarine) 76
V-5 (US submarine) 76
V-6 (US submarine) 76
V-7 (US submarine) 76
V-8 class fleet submarines 76-7
V-8 (US submarine) 77
V-9 (US submarine) 77
V-80 (German submarine) 159
V-300 (German submarine) 159
V class destroyers 11
Vanguard, HMS 93
Versailles, Treaty of 117
Vesuvius, USS 40-3, **41**
    specifications 42
Vian, General Giuseppe 128
Vickers 84
'Victor II' class nuclear submarines 167
Victorious, HMS 150, 154, 155, 158
Vindictive, HMS 111
Virginia, CSS 17-18, 30
Viribus Unitis 65
    specifications 65-6
Viribus Unitis class dreadnoughts *61*, 61-6, *63*, **65**
    specifications 65-6
Vitse Admiral Popov 26-9, *27*
    specifications 28
Vladivostok 37, 51
Volage, HMS 44
Vosper Thornycroft 171, 172

W class destroyers 11
Wa-201 (German submarine) 159
Wa-202 (German submarine) 159
Wadsworth, USS 80
Wainwright, USS 80
Walke, USS 78
War of 1812 11
Ward, USS 81
Warrington, USS 78
Warrior, HMS 12
warship design 9, 21, 28, 56-7, 83, 111
    commercial 171-6
    costs 10, 12, 26, 57, 58, 134-5, 169-70, 171
    design responsibility 13, 23, 25, 176
    factors influencing 10-15
    industrial capability 12-14, 72, 146
    intelligence assessments 11, 48-9
    operational environment 14-15
    and propulsion 14
    technical capability 12-14, 42-3, 60
    threat perception 10-11, 48
    and the Washington Naval Treaty 115
Washington Naval Disarmament Treaty 93, 95, 112, 115, 124, 131, 132, 136, 139, 142
Wasp, USS 145
Watt, Chief Constructor 104
Watts, Sir Philip 31, 57
waves 14
weather 15, 23-4, 27-8, 58
Weehawken, USS 19
Westerwald 120
White, Sir William 45, 46, 47, 48, 49
Wickes class destroyers 81, **82**, 83
    specifications 82
Wilkes, USS 80
William Cramp & Sons 40
Wilson, Sir Arthur 58
Winslow, USS 80
Wk-202 (German submarine) 159
Worden, Lieutenant John Lorimer 18
Worden, USS 176
Wyoming, USS 20

Yakhunin, Y I 177
Yalu River, Battle of 130
Yamato 143, 144-5, **146**, 147
    specifications 146
Yamamoto, Admiral Isoroku 144
Yamato class super battleships 142-7, *143*, *144*, **146**
    specifications 146
'Yankee' class nuclear submarines 165, 167
Yarnall, USS 81
Yarrow Shipbuilders 14, 171, 172
Yellow Sea, Battle of the 50
Yorktown, USS 135, 144
Yugoslav navy 65

Zalinski, Lieutenant 40
Zenker, Admiral Wolfgang 117, 118